HOW TO DESTROY AMERICA IN THREE EASY STEPS

HOW TO DESTROY AMERICA IN THREE EASY STEPS

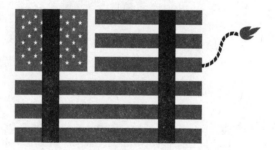

BEN SHAPIRO

BROADSIDE BOOKS
An Imprint of HarperCollins*Publishers*

HarperCollins books may be purchased for educational, business, or sales promotional use. For information, please email the Special Markets Department at SPsales@harpercollins.com.

Broadside Books™ and the Broadside logo are trademarks of HarperCollins Publishers.

FIRST EDITION

Library of Congress Cataloging-in-Publication Data has been applied for.

ISBN 978-0-06-300187-9

20 21 22 23 24   LSC   10 9 8 7 6 5 4 3

To the founders, who created the greatest country in the history of the world; to the Americans who struggled and fought to fulfill the promises they made; and to my children, who inherit the gift of America from all of them.

CONTENTS

INTRODUCTION

W hat holds America together?

That question has, in recent years, taken on renewed urgency. Increasingly, Americans don't like each other. They don't want to associate with one another; they don't want to live next door to one another. More and more, they don't want to share the same country anymore. Red areas are getting redder. Blue areas are getting bluer. According to a November 2018 Axios poll, 54 percent of Republicans believe that the Democratic Party is spiteful, while 61 percent of Democrats believe the Republican Party is racist, bigoted, or sexist. Approximately one-fifth of both Republicans and Democrats consider the opposing party "evil."[1] A Pew Research poll from 2016 found similar numbers: 70 percent of Democrats say Republicans are close-minded, while 52 percent of Republicans say Democrats are close-minded; the same poll found that 58 percent of Republicans had an unfavorable impression of the Democratic Party leading up to the election, while 55 percent of Democrats felt the same.[2]

A 2017 *Washington Post* poll found that seven in ten Americans thought America's political polarization is now as severe as it was during the Vietnam War era, reaching a "dangerous low point."[3] A 2019 survey from American Enterprise Institute found that about half of Americans believe the other party doesn't want what's best for the country. That's likely because Americans increasingly misperceive the nature of those who vote for the opposite political party: both Democrats and Republicans radically overestimate the secularism and radicalism of the constituency of the Democratic Party, for example.[4] According to another study from More in Common, 55 percent of Republicans and Democrats believed that a majority of the opposing party believed extreme views; in reality, that number was 30 percent. So, for example, Democrats believed that only half of Republicans would acknowledge that racism still exists in America; in reality, the number was approximately 80 percent. Conversely, Republicans believed that just half of Democrats were proud to be American; the actual number was about 80 percent.[5]

All of this is having real-life bleed-over effects. According to Pew Research, 79 percent of Americans believe that we have "far too little" or "too little" confidence in each other, and 64 percent believe Americans' level of trust in each other has been shrinking.[6]

The center, philosophically and culturally, isn't holding.

As a matter of historic timing, this polarization is odd. The issues that tore America apart over the past centuries have been radically alleviated. Despite the protestations of the liberal media, racism is at an all-time low in the United States; prosperity was, until the coronavirus pandemic, at an all-time high.

We should be happy together.

And yet, increasingly, Americans seem to be looking for a nonamicable divorce. And both sides want the silverware and the dog. From the Right, the outlook for a united America looks grim: conservatives perceive a triumphalist, aggressive Left, hell-bent on rewriting basic American notions, cramming down an extreme vision of identity politics, cheering the demographic change they insist will inevitably result in a permanent political and cultural ascendancy.

From the Left, the outlook for a united America looks similarly grim: Leftists see a reactionary Right, willing to cut any corner in order to maintain their grip on fading hierarchies of power, clutching at the last vestiges of that old order.

These competing visions have defined the Trump presidency. President Trump represents a sort of political optical illusion: Do you see a blue and black dress, or a white and gold dress? There's no way to see both simultaneously. For those on the Right, Trump represents a seawall against the encroaching, rising tide of radicalism on the Left. His serious character flaws simply become secondary concerns when the future of the nation is at stake. Should Trump lose the presidency in November 2020, conservatives are likely to panic; the potential for national divorce rises dramatically.

For those on the Left, Trump represents confirmation of their worst characterizations of the right: crude, bigoted, and corrupt. The willingness of conservatives to accept Trump, despite all of these flaws, represents further confirmation that the conservative movement was rooted in retrograde impulses papered over with the language of small government. Should Trump win reelection in November 2020, Leftists are likely to panic; the potential for national divorce rises dramatically.

But Trump isn't really the issue, of course. He's merely the symbol of a broader rift in America that predates his presidency, and has been growing, decade by decade.

In order to heal that rift, we must first try to remember why we got married in the first place—and why we've stayed together all these years.

DISINTEGRATION VS. UNION

This is hardly the first time Americans have considered divorce. Indeed, during nearly every major crisis in our history, a contingent of Americans suggest that divorce might be preferable to living together. After all, the logic goes, not all that much holds us together—America is a marriage of interests, not a love match. When the convenience wanes, the marriage ends. Better that we should go our separate ways, or radically redefine Americanism itself—which will end with the same result.

This strain of thought runs from the slaveholding secessionists through the early-twentieth-century political progressives through today's alt-right and identity-politics Left. All of these movements represented a minority of Americans; all had and have outsize influence. The philosophy of division is a philosophy of power politics, a philosophy that paints America as a mythical construct, instituted by those at the top of the hierarchy in order to reinforce their own control. It is a philosophy that derides any notion of American unity as a lie, and bathes that which links us—Abraham Lincoln's "bonds of affection" and "mystic chords of memory"—in acid, disintegrating our ties and casting us all adrift.

Throughout this book, we shall call this strain of thought Disintegrationism.

Then there is another strain of thought. Throughout American history, this strain of thought has emerged victorious—though never without pain and struggle, and sometimes at the cost of death. This philosophy argues that what unites Americans is far stronger and deeper than what divides us, that our vows to one another were cemented in blood, that we are inextricably intertwined. A separation would kill us both.

This strain of thought runs from the founding fathers through

Abraham Lincoln through the civil rights movement. This strain of thought championed reason and universal morality above passion and tribalism, and emerged with a belief in the value of democracy and individual rights—principles that were always true, but never properly applied. This strain of thought suggests that America is always an imperfect union, but it is indeed a union—and that we are always in the process of strengthening and growing that union, built on the foundations of founding ideals.

Throughout this book, we shall call this strain of thought Unionism.

Most Americans are Unionists. But they are under attack: steady, unyielding attack by those who support Disintegrationism. Our bonds are fraying. What is left is chaos. Without the ties of Unionism, the center cannot hold. And it isn't.

THE ELEMENTS OF UNIONISM

So, let's get more specific. What, exactly, has allowed America to stay a country? And why should we continue to do so today?

There are three elements that make America *America*.

First, American philosophy.

The philosophy of the United States rests on three basic principles: first, the reality of natural rights, which preexist government, inalienable and precious; second, the equality of all human beings before the law, and in their rights; and finally, the belief that government exists only to protect natural rights and to enforce equality before the law. American philosophy believes these propositions are "self-evident," in the words of the Declaration of Independence. The founders attempted to implement American philosophy through a unique set of institutions. The Constitution of the United States was a compromise document, designed to enshrine American philosophy via a limited

government system. That constitutional system's enumerated powers
balanced the necessity for action embodied in the legislative power
with the necessity to avoid tyranny; the constitutional system's checks
effectively balanced the requirement of an executive powerful enough
to respond to threats and enforce law with the requirement to avoid
despotism embodied in checks and balances; the constitutional system's federalism was constructed to frustrate national schemes to
subsume the character of local communities, while simultaneously
preventing those local communities from becoming autocracies.

Next, there is American culture. That culture is characterized by
four distinct elements. First, a tough-minded tolerance for the rights
of others, particularly when we don't like how others exercise their
rights—we have to agree to disagree, and to get over it. Second, our
culture prizes and cherishes robust social institutions, which create a
social fabric that allows us to trust one another in the absence of compulsion from government. Third, American culture has always carried a rowdy streak in defense of liberty: we must be willing to stand
up for our freedom and that of others. Finally, American culture has
always celebrated and rewarded those with a sense of adventure—the
pioneers, the cowboys, the inventors, the risk takers.

These attitudes spring from our philosophy, and have seeped into
every aspect of American life and thinking. Americans need not be familiar with founding philosophy to rely on our rights, or to be familiar
with their magnetizing attraction. The ringing statements of the Bill
of Rights infuse our language; our arguments over politics inevitably
sound in the context of free speech and protection of private property;
our arguments over government routinely invoke due process protections and protection against unreasonable government intrusion; our
arguments over social issues center on freedom of religion and freedom of association. Our duties, by contrast, spring from traditional
understanding of the social fabric, rooted in Judeo-Christian values.

Finally, there is American history. American history has traditionally been read as a story of ever-improving fulfillment of American
philosophy and culture through proper exercise of American institutions. Traditionally, Americans have learned that the values of the
Declaration of Independence are eternal and true; that our culture of
rights has been broadened in application over time by heroic struggle and through horrific pain; and that the constitutional system represents liberty, increasingly effectuated. American history, then, is a
story of triumph of freedom over the tragedy of human nature, the
victory of liberty over slavery and bigotry.

These three elements—America's philosophy of reason, equality, liberty, and limited government; America's culture of individual
rights and social duties; and America's shared history—define our
country. No single one of these elements is sufficient to bind America. America's philosophy alone, without shared culture and history,
is sterile and impractical: the philosophy must be combined with the
shared living of culture and shared memory of history, or else remain
empty. Americans don't feel the swelling of brotherhood in reason
alone. America's culture of rights, without philosophy and history,
falls prey to the passions of the crowd: if rights prevent the mob from
doing its will, rights quickly dissolve. America's history, without a
philosophy of reason and a culture of rights, breaks down into a series
of disconnected events, motivated by sheer power politics.

What's more, one missing element leaves America in dire straits.
Without America's philosophy, reason collapses into tribalism; without America's culture, individual rights collapse into collectivist tyranny or duties collapse into libertinism; without American history, the
symbols that unite us divide us.

Americans, Abraham Lincoln stated in his First Inaugural Address, must stick together—we had to live together, or die alone. "We
are not enemies, but friends," Lincoln stated, just weeks before the

Confederacy fired on Fort Sumter. "We must not be enemies. Though passion may have strained it must not break our bonds of affection."

THE ELEMENTS OF DISINTEGRATIONISM

But there is another story about America. And this story about America is gaining ground, particularly on the political left.

Two weeks before the 2020 Iowa caucuses, Senator Bernie Sanders, Independent of Vermont, the intellectual thought leader and emotional avatar of young Democrats across America, gave a rally in Ames. The lines stretched around the block, thousands strong.[7] The event opened with the popular rock band Portugal. The Man; one of the musicians, Zack Carothers, then got up and ushered onstage three Native American women, explaining, "the land that we are on is not ours." The women then called for "land reparations" and explained that Iowa had been stolen from indigenous tribes. After that, radical filmmaker Michael Moore took the stage to explain that America was built "on genocide and built on the backs of slaves," that American racism had not abated, and added that America was a "system set up to benefit the few at the expense of the many." Following Moore, Democratic representative Alexandria Ocasio-Cortez of New York, a self-described democratic socialist, told a screaming throng that they were part of a "movement for social, economic and racial justice . . . a movement to transform our public policy so the United States can . . . finally advance 21st century human rights." Ocasio-Cortez proclaimed, "We need fundamental change in the United States of America. . . . It's going to require us to transform and grow as individuals." She called for a "collaborative movement . . . to build a more advanced nation."

Finally, Bernie Sanders took the stage. After repeating his litany of complaints about President Donald Trump, he pledged to fulfill

Ocasio-Cortez's promises. Sanders ripped into a supposed oligarchy impoverishing Americans, the rigged system of the United States. "We want a nation that works for all of us, not just the few," said Sanders. And then he, too, pledged fundamental change: "All of us, together, rolling up our sleeves, standing up and fighting to create the kind of nation that you and I know we can become." He then unspooled a list of policy proposals ranging from nationalized health care to government relief of student debt, from massive government control of the energy sector to extraordinary tax increases.[8]

The America of the past had to be left behind. American philosophy was corrupt and exploitative; American culture was racist and cruel; American history was a litany of abuses, punctuated only by sporadic revolutions directed at overthrowing her philosophy and culture.

This is the Disintegrationist view, boiled down to bumper stickers.

The Disintegrationist view launches direct, unyielding attacks on American philosophy, culture, and history.

American philosophy is under attack, with Disintegrationists claiming that natural rights do not exist—that no rights are discoverable from human nature and reason, because neither human nature nor reason exists. Human nature is inherently malleable, and reason a mere tool of power, wielded by political enemies in order to suppress dissent.

Likewise, equality before law is morally wrong, according to Disintegrationists—such equality merely reinforces preexisting hierarchies of power. Instead of equality before the law, or equality in individual rights, Disintegrationists seek equality of outcome.

Finally, Disintegrationists see government not as a guarantor of individual rights and equality before law, but as an overarching cure-all, available to change the hearts and minds of men. To that effect, the institutional framework installed by the founding fathers in order to effectuate their philosophy has come under assault by Disintegrationists, too. Disintegrationists oppose the doctrine of enumerated powers as insufficient to meet the needs of citizens; they oppose checks and balances as barriers to progress; they oppose federalism as a framework for oppression.

American culture is also under attack, with Disintegrationists claiming that rights themselves are a threat to the common good. Free speech must be replaced by hate speech regulations, with hate itself left undefined. Freedom of religion must be replaced by secular universalism. Freedom of association and contract must be prohibited, so long as that freedom cuts against the appropriate standards of ethnic, racial, or sexual diversity (under this standard, for example, an all-black school is considered diverse, while a police department that doesn't represent ethnic populations proportionately is considered discriminatory, even if that police department staffs based on meritocratic concerns). Due process must be supplanted with mob rule, private property with public need.

Disintegrationist culture makes the further claim that social institutions have buttressed America's evils. Those institutions, in the view of Disintegrationists, must be leveled in order to build a better world. The village must be burned down, and a glorious new city built in its place.

Disintegrationist culture claims, too, that Americans' stubborn willingness to defend their rights represents a pigheaded defense of a corrupt and hierarchical system; Americans must be trained to accept diktats from their government before they can be cured of their individual heresies.

Disintegrationist culture seeks to replace America's love of risk-taking with a sense of solicitousness from the collective. Those who take risks are to be treated as greedy leeches, and any system that rewards risk-taking must be treated as morally repugnant. Instead, Americans should cultivate a sense of dispossession. Only through rent-seeking can ultimate justice be done.

Finally, America's history is under severe threat. The Disintegrationists claim that America's traditional history is a myth: that the *true* story of America is a story of exploitation, that the ideals of the Declaration of Independence were a self-flattering parody when written, that the Constitution of the United States was meant to enshrine power hierarchies, as well as bigotry of all forms. America has been an imperialist monster hell-bent on world domination, a propagator of rapacious capitalism, a faux democracy. In this view, there is no history to bind us—in fact, history separates us. The American flag itself represents nothing more than a cynical joke, in the Disintegrationist view.

Now, as we'll explore, the Disintegrationist view isn't merely wrong, it's dangerous. But it's gaining steam—day by day, hour by hour. And when the Disintegrationist view becomes the majority view in the United States, the United States will no longer be united.

THE DISINTEGRATIONIST PROBLEM

In order to argue that America's philosophy is wrongheaded, her culture diseased, and her history evil—to treat America as the great exploiter rather than the great liberator, wealth creator, and rights defender—Disintegrationists must engage in an extraordinarily selective reading of reality. They must home in in excruciating detail on America's sins, which, in context, would be fine—but rob that history of all context or subsequent history. Exploitation is a feature of every human society, and repeated mistreatment by some groups of other groups is a similarly common feature. What is *uncommon*—indeed, unprecedented in human history—are *prosperity, peace,* and *freedom.*

It is simply undeniable that capitalism, founded on protection of property rights—the ideology of the founding fathers—has been

uniquely successful in spreading peace and prosperity both domestically and around the globe. The simple fact of the matter is that
since the dawn of the Enlightenment, the enshrinement of individual
rights, and the advent of protection for private property—the roots
of capitalism—global GDP has increased exponentially, in shocking
fashion. In the year 1 BCE, global GDP amounted to $183 billion; in
1000, global GDP amounted to approximately $210 billion; in 1500,
it was still just $431 billion; in 1700, $643 billion; as of 2013, $101 tril-
lion. That is a 15 percent increase in the first millennium; a 105 percent
increase from 1000 to 1500; a 49 percent increase between 1500 and
1700; and a 15,700 percent increase from 1700 to present.[9]

It is similarly undeniable that the spread of peace has been a direct
result of American hegemony. On a year-by-year basis, international
war deaths have decreased precipitously since World War II, from a
high of nearly 200 deaths per 100,000 people at the end of that conflict to a low of well below 0.5 deaths per 100,000 people at the turn
of the twentieth century.[10] Global life expectancy has doubled since
1900.[11] Furthermore, America has become the most tolerant country
on earth. According to the *Washington Post*, a new Swedish survey
found that people from the United Kingdom, America, Canada, and
Australia, as well as certain Latin American countries, were "most
likely to embrace a racially diverse neighbor." Other European countries aren't nearly as tolerant.[12] And none of those countries has ever
elected a black man—twice—with more than 65 million votes each
time, to serve as the leader of those countries.

Finally, it is perfectly obvious that global freedom has expanded
wherever American influence has expanded. According to the Pol-
ity Project at the University of Maryland, democracy is actually
at a global high, and has been in particularly steep ascent since the
death of the Soviet Union[13]—a collapse brought about, of course, by
America's willingness to "pay any price, bear any burden, meet any

hardship, support any friend, oppose any foe, in order to assure the survival and the success of liberty," as John F. Kennedy put it in his Inaugural Address. The oft-cited bumper sticker history that America freed Europe of tyranny twice, ended the scourge of communism, and has liberated billions around the globe happens to be true.

The bar for Disintegrationists, then, is high.

To surpass it, Disintegrationists must employ a clever, seductive, and deeply vicious strategy.

THE STRATEGY OF DISINTEGRATIONISM

So, how have Disintegrationists succeeded in convincing millions of Americans that America's philosophy, culture, institutions, and history are all worth overthrowing? They've provided a subversive but seductive view of America as an evil actor—and provided an alternative Unionism rooted in intersectional solidarity. Intersectionality, in its original iteration, was perfectly plausible: it suggested that Americans may be targeted based on membership in more than one minority category. So, for example, a black woman might meet discrimination in a different way than a white woman. But intersectionality has instead become a rallying cry for Disintegrationists who aver that America is subject to unbending, rigid hierarchies that can be torn down only by uprooting the entire American system. Thus, membership in a historically victimized group serves as the glue to hold together an anti-Unionism coalition. In opposition to the system lies a new solidarity.

This argument is extremely seductive, especially given the underlying philosophical lie of the political left: that all disparity represents a form of discrimination. Since disparities have existed between all groups at all times, such disparities cannot disappear. But by convincing Americans that any unexplained disparity is the result of the

American system—philosophy, culture, institutions, and history—
Disintegrationists have a succinct and irrefutable argument in favor
of tearing down the system. Any evidence of disparate treatment becomes an argument against Unionism.

This is an emotionally resonant pitch. Traditional Americanism
suggests that while our system has never been perfect, it has grown
increasingly so—and this means that it should be easier to succeed
today, without the obstacles of bigotry that have plagued our history,
than ever before. That worldview places an awful responsibility on individuals: if you fail to succeed, you can certainly blame personal dis-
advantages, but it becomes difficult to blame a miasmatic, existential,
systemic, flag-draped boogeyman haunting your dreams. Additional
freedom means additional responsibility.

If, however, all disparity can be chalked up to the system, then
personal responsibility becomes a secondary concern. Failures are
no longer individual, but systemic. In fact, every failure becomes an
additional brick in the wall of evidence against America. The only
solution, again, rests in coalitional politics designed to rewrite the
American bargain.

This outlook has become a rote part of radical Democratic politics:
the notion that a coalition of the supposedly oppressed must rise up
and rewrite the entire nature of the American bargain. Thus Senator
Kirsten Gillibrand announces that "resistance is female, intersectional
and powered by our belief in one another." Social media–savvy Democratic presidential candidates strive to demonstrate their own intersectional bona fides, lest they be labeled patronizing fellow-travelers;
stories of victimhood at the hands of the country that made them
wealthy and prominent abound. And these same candidates aren't
content to merely criticize specific policies of the government, which
would be the root of all politics. Instead, they criticize the "system"
itself. Thus Democratic senator Elizabeth Warren of Massachusetts

terms the criminal justice system "racist"; Senator Kamala Harris of California explains that identity politics shouldn't be eschewed—in fact, she argued, the phrase itself was designed to "minimize and marginalize issues that impact all of us. It's used to try and shut us up."[14] As former Texas Democratic representative Beto O'Rourke, the id of the Democratic Party, put it in 2019, "this is a country that has been defined by foundational systemic endemic racism since the very founding of this country."[15]

Thanks to the politics of Disintegrationism, the media routinely focus on demographic change in the United States as a harbinger of radical political change—the suggestion being, of course, that demographics is destiny. Steve Phillips of the Center for American Progress triumphally wrote in *The Nation* that "the single greatest force shaping American politics today is the demographic revolution that is transforming the racial composition of the US population." According to Phillips, the Democratic Party's hopes lie in more race-based thinking: "The concerns of people of color should be driving politics today and into the future."[16] This is hardly fringe thinking in Democratic circles; for a large swath of the Democratic intelligentsia, it's a strategic linchpin. It's no small irony that alt-right racists hold the precise same viewpoint—that the changing ethnic composition of America represents an inherent threat to the American system—and prescribe their own form of racial Disintegrationism as a solution. Unfortunately, that alt-right mind-set has, from time to time, crossed over into the Republican Party, too.

All of this is deeply divisive. But in the Disintegrationist view, the *true* anti-unity forces lie among advocates for traditional Americanism. Unity, say the Disintegrationists, can be born of opposition to the system. But since the system itself is inherently divisive, standing up for it is *ackshually* the reason for our political and cultural fragmentation. This is why Disintegrationists have categorized culturally unifying symbols like the American flag as inherently divisive. If America, as a concept, is polarizing, so too is the flag.[17] Kneeling for the national anthem *ackshually* represents more authentic unifying behavior than standing for it (Beto O'Rourke suggested, "I can think of nothing more American than to peacefully stand up, or take a knee, for your rights, anytime, anywhere, in any place").[18] America's traditional reliance on reasoned conversation is itself deemed polarizing—after all, civility was a tool of the Confederacy (in point of fact, the Confederacy was not actually well known for its civility, as the 600,000 American corpses of the Civil War should amply demonstrate).[19] Scientific investigation is deemed bigoted, and meritocracy itself derided as discriminatory. Belief in free markets—even opposition to nationalized health care—is evidence of America's roots in slavery.[20] Adherence to American institutions like federalism and the Electoral College is castigated as inherently discriminatory.[21]

The system itself is oppressive and repressive, and must be fought. Anyone who believes that America's philosophy, culture, institutions, and history are unparalleled as a source of good and rights is derided as a Pollyannish jingoist at best, a bullheaded bigot at worst. There is a reason that when I spoke at Boston University on this topic—the lecture was titled "America Was Not Built on Slavery, It Was Built on Freedom"—hundreds of protesters gathered outside, immediately labeling me racist. Never mind that the speech focused specifically on the horrors of slavery and Jim Crow, as well as the heroic freedom struggles of black Americans: to even point out that the freedom fighters won, that they were fulfilling the promises of the founders, that they were cashing Martin Luther King Jr.'s famous "promissory note" guaranteed in the Declaration of Independence—all of this was part and parcel of a broader racist viewpoint to be discarded.

Thus, those factors that should unify us are *themselves* portrayed as divisive. And only kowtowing to the Disintegrationist worldview earns moral absolution.

And Disintegrationists demand such kowtowing.

Disintegrationists leverage the power of cultural institutions to target and destroy those who stand in their way. Disintegrationist strategy relies on a simple rule: the squeaky wheel gets the grease. In the world of social media, this means targeting corporations who are too risk averse to stand up for either free speech, or for their own values. Thus Disintegrationist activists spend their days trolling conservative shows for clips to misrepresent, then crowdsourcing support on Twitter for boycotts that never actually materialize. A few tweets are often enough to prompt advertisers to drop hosts, prompt social media platforms to demonetize or de-platform supposed violators, or even shift entire broad-based policies.

The ever-shifting boundaries of political correctness, as enforced by the Disintegrationist elitist mob, force silence or compliance.

Corporations—which, after all, are motivated by profit rather than principle—often cave to even mild pressure. Corporations are nonideological in nature, contrary to left-wing characterization; they're legal structures designed to limit liability. This means that a few interns mouthing off at a massive company can shape policy from the inside—and it also means that a mild media firestorm may be enough to shift even billion-dollar companies toward more restrictive speech standards.

Even charities aren't spared the Disintegrationist lash. The Boy Scouts can be forced to denounce its own principles, should those principles comply with traditional attitudes toward Judeo-Christian morality. Chick-fil-A can be forced by both governmental and nongovernmental pressure to drop charitable giving to organizations like the Salvation Army.

College campuses, where Disintegrationists effectively control many administrations, have become the scene for egregious de-platformings and Maoist struggle sessions. Students are taught to confess their "white privilege"—a key component of the intersectional theory undercutting traditional American notions of individual responsibility and rights—and if they do not, they are labeled bigots. Professors are forced to abandon campus if they refuse to kneel before the rule of the Disintegrationists. Scientists are forced to abandon their studies if those studies provide unpopular results.

Hollywood, too, has become the tip of the sphere in the intimidation business. Because celebrities are particularly vulnerable to bad press, they are often forced to comply with the Disintegrationist viewpoint or face firings. The mildest objection to the most radical leftist agenda items earns a career-threatening rebuke. Mario Lopez, for example, committed the grave sin of suggesting that it would be "dangerous as a parent" to determine based on a three-year-old's gender confusion that the child is indeed suffering from gender dysphoria, adding "You're just a kid." This was enough to spur E! to pressure Lopez for a public apology, which he duly gave.[22]

The man has a family to support.

So do we all.

The Disintegrationists know this. And they are unwilling to leave any area of American life untouched by their acidity. Sports must be turned into referenda on the intolerance of biology; musicians must be forced to comment on the latest social justice craze.

THE ENDGAME

Every piece of connective tissue in American life is being stripped away by Disintegrationists. This is happening constantly, all around

us. We cannot watch a football game without confronting Disintegrationism; we cannot pick up a celebrity magazine, attend a church
event, or go to a restaurant without first considering politics.

This is effective.

It's effective because it's exhausting. It is meant to be. It is meant to
convince Americans to throw up their hands and simply comply with
the dictates of the Disintegrationists, or to force Americans to divide
up every element of daily life politically. Either solution is unwork-
able for a country that wishes to remain united. Unionism suggests
that despite our differences, we are, at root, Americans. Disintegrationism suggests that despite our American passports, we are, at root,
different.

The dirty little secret of Disintegrationism is that there is no unity
at the end of the destruction. There is only more destruction. The
revolution will eat its originators. No new world will be rebuilt after
the razing of America's philosophy, culture, institutions, and history.
Tribalism will simply replace national unity. The glue of opposition
that currently binds together the disparate factions of Disintegration
will simply melt away. All that will be left are polarized groups, seeking their own interests.

That means that the fight to save America is, first and foremost, a
fight to defend Unionism. It is to that fight we now turn.

HOW TO
DESTROY
AMERICA IN
THREE EASY
STEPS

CHAPTER 1

THE AMERICAN PHILOSOPHY

The American creed is simple.

"We hold these truths to be self-evident, that all men are created equal, that they are endowed by their Creator with certain unalienable Rights, that among these are Life, Liberty and the pursuit of Happiness; that to secure these rights, Governments are instituted among Men, deriving their just powers from the consent of the governed."

Our founding fathers pledged their lives, their duty, and their sacred honor to these principles. So did the soldiers who fell at Gettysburg, the civil rights heroes who stood for their fulfillment, the men and women who brought freedom to billions across the globe. The Declaration's ringing endorsement of equality, individual rights, and democracy has lost none of its power, despite being the subject of cliché, or despite deliberate misinterpretation.

The words of the Declaration of Independence still represent a singular, bedrock viewpoint. More than that, the theme of the Declaration represents a continuous through-line in American thinking. Ex-slave and second founding father Frederick Douglass found in the Declaration of Independence the halcyon call to freedom—Douglass endorsed its "great principles of political freedom and natural justice"—and a brutal rebuke to the toleration of slavery in the United States.[1] Abraham Lincoln saw in the Declaration of Independence the philosophical *axis mundi* of the United States:

> Wise statesmen as they were, they knew the tendency of prosperity to breed tyrants, and so they established these great self-evident truths, that when in the distant future some man, some faction, some interest, should set up the doctrine that none but rich men, or none but white men, were entitled to life, liberty and pursuit of happiness, their posterity might look up again to the Declaration of Independence and take courage to renew the battle which their fathers began—so that truth, and justice, and mercy, and all the humane and Christian virtues might not be extinguished from the land; so that no man would hereafter dare to limit and circumscribe the great principles on which the temple of liberty was being built.[2]

Martin Luther King Jr., more than a century later, saw in the Declaration of Independence a promissory note yet to be fulfilled. From the steps of the Lincoln Memorial—the site built to honor the president who died in his own attempt to fulfill the words of the Declaration—King called for the completion of that task. "This [promissory] note was a promise that all men, yes, black men as well as white men, would be guaranteed the unalienable rights of life, liberty, and the pursuit of happiness," King thundered. "I have a dream that one day this nation will rise up and live out the true meaning of its creed, 'We hold these truths to be self-evident, that all men are created equal.'"[3]

The words were not cataclysmically new, nor were they meant to be. They were meant to embody very *old* principles—principles rooted in Greek and Roman philosophy, in Judeo-Christian values, in English

tradition. As Thomas Jefferson would write in May 1825, a year before his death, the goal was to embody eternal ideals: "Not to find
out new principles, or new arguments, never before thought of, not
merely to say things which had never been said before; but to place
before mankind the common sense of the subject, in terms so plain
and firm as to command their assent, and to justify ourselves in the independent stand we are compelled to take." The Declaration of Independence aimed at providing an "expression of the American mind,"
synthesizing "sentiments of the day" drawn from both contemporaneous writing as well as "elementary books of public right, as Aristotle,
Cicero, Locke, Sidney . . ."[4]

So, what exactly was that "expression of the American mind"? Why
does it remain relevant today?

The philosophy of the United States centers on three central principles, as articulated in the Declaration: on the reality of natural rights
to life, liberty, and the pursuit of happiness that preexist government;
on the equality of men before the law; and on the notion that government is instituted only to protect those preexisting rights and equality
of men before the law.

The philosophy of the Declaration of Independence remains the
ideological glue that unifies us. Without it, as the Civil War demonstrated, we collapse. It is a philosophy of high ideals—unreachable
ideals, perhaps, but eternal ones. We abandon it at our peril.

WE HAVE INDIVIDUAL, NATURAL RIGHTS

The Declaration of Independence posits that rights are self-evident,
and that all individuals are bestowed those rights by Nature and Nature's God. But what are our rights, and how can we know them?
Why should human beings have rights?

These questions, as we will soon see, are quite open. They stand at the center of our current political conflicts. Some suggest that rights come from government, and that government therefore has the power to remove such rights. Some suggest that rights are merely utilitarian constructs—that rights are a handy way of thinking about "stuff other people shouldn't do to us," but can be overridden if the concerns of the many outweigh the concerns of the few.

The founding fathers, however, based their belief in rights in the framework of *natural rights*. Why? Because rights that spring from human beings aren't rights at all—they're privileges that may be dispensed with at any time. Rights that emanate from a higher source are indeed *unalienable*: they cannot be given away, taken away, or infringed upon, because they come from a source higher than power alone.

This philosophy did not spring full-blown to the minds of the founders. It had a long and rich history. The birth of natural rights came, in turn, from natural law philosophy. Natural law philosophy was rooted in two traditions that gradually fused into the belief system of the founders: Judeo-Christian morality, or Jerusalem, and Greek reason, or Athens. Judeo-Christian morality posited that human beings had inherent value; the Bible's ringing statement of human equality in Genesis 1:27 provided fruitful ground for the most basic rebuke of caste systems: "And God created man in His image; in the image of God He created him; male and female He created them." This language served as a direct contrast to other ancient civilizations, which invested leaders alone with the image of God.[5]

Meanwhile, Greek philosophy suggested that what made man unique was his ability to use reason. Human beings, the Greeks believed, were not innately malleable or changeable. We were crafted by nature to use our minds; it was that capacity that separated us from the animals. Plato asserted that the purpose of the human soul was "managing, ruling, and deliberating."[6] Aristotle suggested that "the work of a human being is an activity of soul in accord with reason."[7]

When the key Judeo-Christian premise—the inherent value of mankind—was added to the Greek belief that what made human beings different was their ability to reason, the concept of natural law was born. The philosophies of Jerusalem and Athens became the poles of Western civilization; at some times, Jerusalem reigned victorious, at other times, Athens. For centuries, Jerusalem and Athens actually pulled in the same direction. When Greek thought was revivified in the Christian West at the beginning of the second millennium, it led to an explosion in learning, particularly at universities sponsored by the Catholic Church. Famed theologian Hugh of Saint-Victor (1096–1141 CE) stated, "Learn everything, later you will see that nothing is superfluous"[8]; Thomas Aquinas (1225–1274 CE) explained, "An error concerning the [science of creation] ends as false thinking about God."[9] Religious conflict didn't tarnish the draw of natural law—it burnished it, providing the basis for a universalism absent in religious sectarianism. No wonder Hugo Grotius, the famous philosopher and jurist, appealed to natural law in the aftermath of the barbarous Thirty Years War, and in the aftermath of the transformative Peace of Westphalia, praising the "dictate of right reason which points out that an act, according as it is or is not in conformity with rational nature, has in it a quality of moral baseness or moral necessity; and that in consequence, such an act is either forbidden or enjoined by the author of nature, God."[10]

Natural law thinking led to the development of natural rights. If human beings had a nature, and if that nature was reason, then man had a right to exercise his reason—and to exercise his will so long as he hurt no one else. As Hugo Grotius stated, "God created man free and *sui iuris* [under his own dominion], so that the actions of each individual and the use of his possessions were made subject not to another's will but to his own."[11]

Natural rights thinking reached its apex in the writings of John Locke (1632–1704). Like Grotius, Locke believed that because human beings have the ability to reason—because that is our *nature*—and because we all have inestimable value as creations of God, we have rights. Locke argued that in a state of nature, human beings had certain rights, rights that preexisted government. As he stated, "The natural liberty of man is to be free from any superior power on earth, and not to be under the will or legislative authority of man, but to have only the law of nature for his rule."[12] So, what were man's rights under the laws of nature? The rights to life, liberty, and property. Human beings have a right to self-defense inherent in their very survival; they have a right to liberty of thought and action so far as they do not violate the rights of others; and they have a right to the property they accrue through their labor.[13] Adam Smith, widely considered the philosophical expositor of free markets, took this view as well:

> [T]he obvious and simple system of natural liberty establishes itself of its own accord. Every man, as long as he does not violate the laws of justice, is left perfectly free to pursue his own interest his own way, and to bring both his industry and capital into competition with those of any other man, or order of men. . . . [14]

Man's nature was to reason; man was entitled to the freedom to use his reason in accordance with that nature. All of the founders believed this. Thomas Jefferson felt so strongly about the value of reason that he urged his nephew to "Question with boldness even the existence of God; because, if there is one, he must more approve of the homage of reason, than that of blindfolded fear."[15] Jefferson hoped that America would establish the fact "that man may be governed by reason and truth." Alexander Hamilton, one of the writers of *The Federalist Papers*, wrote similarly in 1775 that God had given existence to man, as well as the means of "preserving and beatifying that existence": the use of "rational faculties."[16] Reason—and the rational understanding of the boundaries of fixed human nature—lay at the root of the revolution. The founders explicitly refused to ground natural rights merely on convention or tradition. As Richard Henry Lee of Virginia, the founding father who brought forward the Lee Resolution setting in motion the Declaration itself, stated, "Rights are built on a fourfold foundation—on Nature, on the British constitution, on charters, and on immemorial usage," but he pushed for an American understanding that rights were primarily rooted in "the broadest Bottom, the Ground of Nature."[17]

This is why the Declaration of Independence sources rights to the "laws of Nature and Nature's God." Human reason, so the theory goes, can examine the character and activities of mankind and determine what mankind was *meant* to do—and what liberties mankind was handed in order to pursue those purposes. Thus, the natural law logic that prizes reason as the highest end of man results in liberty as the chief requirement of man. Man's unique nature lies at the center of any rationale for human rights.

Those natural rights were considered inviolable by the founding fathers. And they lay the groundwork for the revolution by requiring that government be a protector of rights rather than the source of them.

ALL HUMAN BEINGS ARE EQUAL BEFORE THE LAW

The founders didn't merely believe that human beings were endowed with rights by virtue of their nature. They believed that all human beings, regardless of their natural talents, were entitled to equality in their rights. Jefferson's original wording in the Declaration of Independence was more exact than the wording with which we are now familiar: he suggested, "We hold these truths to be sacred & undeniable; that all men are created equal & independent, that from that equal creation they derive rights inherent & inalienable."[18] Equal and *independent*. Equal in *creation*.

This is, to be clear, not entirely obvious. To declare, as the Declaration of Independence does, that all men are created equal in their rights seems incredibly counterintuitive. After all, we may all be human beings subject to similar dictates of human nature, but we are deeply unequal in a variety of other ways: some of us are smart, some foolish; some tall, some short; some athletic, some nerds. So how could the founders suggest that it is a self-evident truth that all men are created equal in our rights?

For most of human history, the wisest of men suggested precisely the opposite. Plato, for example, posited that in order to maximize the possibility of philosophers guiding society, society ought to be ordered from the top down; he recommended that "the philosophers rule as kings," that the community be rigidly ordered into castes of workers, warriors, and philosophers. Aristotle promulgated a rather similarly striated notion of human inequality, arguing that the "relation of male to female is by nature a relation of superior to inferior and ruler to ruled. The same must of necessity hold in the case of human beings generally." This, Aristotle argued, justified natural slavery: "those who are as different from other men as the soul from the body or man from beast . . . are slaves by nature."[19]

It is here that the Judeo-Christian values of the founding become clearest. The Bible makes clear that in God's eyes, we are all valued equally. Leviticus 19:15 dictates, "You must not pervert justice; you must not show partiality to the poor or favoritism to the rich; you are to judge your neighbor fairly." That theme is made more explicit in Christian theology: Galatians 3:28 dictates, "There is neither Jew nor

Gentile, neither slave nor free, nor is there male and female, for you are all one in Christ Jesus." Human equality is the presupposition of Judeo-Christian thinking.

Likewise, British tradition, which undergirded American philosophy, supported the notion of equality in citizenship. The Magna Carta expressly suggested notions of equality before the law and due process of law. It is no shock that the *Journal of the Proceedings of the Congress* in 1774 carried on its title page a picture of twelve arms grasping a document titled "Magna Carta."[20]

The founders found in Judeo-Christian tradition, English constitutionalism, and the dictates of reason a wholesale rejection of the ancients' focus on human inequality. They relied heavily on the philosophy of John Locke, who stated that in the state of nature, natural law governed, rooted in reason—and that natural law dictated human equality. Locke explained in his *Second Treatise of Government*:

> The state of nature has a law of nature to govern it, which obliges every one: and reason, which is that law, teaches all mankind, who will but consult it, that being all equal and independent, no one ought to harm another in his life, health, liberty, or possessions. . . .[21]

The founders embraced Locke's philosophy on human equality. In his influential *Common Sense*, probably the most widely read political pamphlet of the founding era, Thomas Paine argued that the distinction of men into "kings and subjects" should have been rejected on the grounds that humans were "originally equals in the order of creation." Paine humorously noted, "One of the strongest *natural* proofs of the folly of hereditary right in kings, is, that nature disapproves it, otherwise she would not so frequently turn it into ridicule by giving mankind *an ass for a lion*."[22] James Wilson, a signatory to both the Declaration and the Constitution, suggested that diversity in the skills, virtues, and dispositions of human beings did not in any way undermine the fundamental claim of human equality: equality in rights, Wilson wrote, "forms a part of that great system, whose greatest interest and happiness are intended by all the laws of God and nature."[23] The Virginia Declaration of Rights, written by George Mason, predated the Declaration by less than a month—Jefferson used it as the basis for the Declaration—and claimed, "all men are by nature equally free and independent."[24] John Adams's Massachusetts Constitution explicitly mimicked the verbiage of the Declaration, with substantially more specificity: "All men are born free and equal, and have certain natural, essential, and unalienable rights."[25]

None of this philosophizing ended the hypocrisy of a society that continued to tolerate slavery, of course. And this didn't go without comment—Dr. Samuel Johnson famously quipped, "How is it that we hear the loudest yelps for liberty among the drivers of Negroes?" But it is worth noting that when the Declaration of Independence was written, slavery was common practice across the world; as we will discuss later, slavery itself was not outlawed in Britain's territories until 1833, and not in India until 1843. And as we will also see, the founders were well aware of the conflict between their ideals of equality and the institution of slavery. None of them claimed that blacks were not, in fact, human beings covered by the Declaration of Independence's high ideals. In fact, as C. Bradley Thompson points out, the argument for the race-based inapplicability of the Declaration's guarantee of human equality didn't emerge in the United States until the militant proslavery movement got under way in the 1830s, in direct response to the abolitionist movement.[26]

The Lockean tradition was carried forward through American history. Abraham Lincoln homed in on the key message of the Declaration in his Gettysburg Address: he stated that the United States had been "dedicated to the proposition that all men are created equal."

Reacting to the incorrect and morally abominable *Dred Scott* decision by the Supreme Court, which declared that black Americans were not protected by the Constitution of the United States, Lincoln explained that the founders were not arguing that all men were created "equal *in all respects*"—in "color, size, intellect, moral development, or social capacity." Instead, the founders meant that all men were equal in "certain inalienable rights, among which are life, liberty, and the pursuit of happiness." Lincoln stated, "This they said, and this meant." The founders understood the hypocrisy of declaring all men equal in a time of slavery; they understood they were not freeing black slaves. "They meant simply to declare the *right*, so that the *enforcement* of it might follow as fast as circumstances should permit," said Lincoln. What, Lincoln asked, did the founders seek to accomplish with the language of the Declaration?

> They meant to set up a standard maxim for a free society, which should be familiar to all, and revered by all; constantly looked to, constantly labored for, and even though never perfectly attained, constantly approximated, and thereby constantly spreading and deepening its influence, and augmenting the happiness and value of life to all people of all colors everywhere. . . . Its authors meant it to be, thank God, it is now proving itself, a stumbling block to those who in after times might seek to turn a free people back into the hateful paths of despotism.

To read the document as a race-specific document would be to shear the Declaration of "its vitality" and "its practical value," and to remove "the *germ* or even the *suggestion* of the individual rights of man."[27]

Of course, civil rights heroes from Frederick Douglass to Martin Luther King Jr. embraced the same interpretation of the equality of mankind. At the fourteenth meeting of the National Negro Business

League in 1913, Booker T. Washington paid tribute to the fiftieth anniversary of the freeing of the slaves by praising "that immortal document, the Declaration of Independence." He noted, "Whether the American negro was meant at the time to be included within the scope and meaning of the words of the Declaration of Independence has been a debatable question." But, said Washington, black Americans would claim their membership in the American creed, so "that in all the future no one will dare question our right to be included in any declaration that relates to any portion of the body politic."[28]

GOVERNMENTS SHOULD PROTECT PREEXISTING RIGHTS AND EQUALITY BEFORE LAW

In the natural rights model, the locus of power lies in the individual, who becomes the repository of indefeasible rights. Government is merely the delegated guardian of individual rights; government has no power to invade those rights, lest it lose its legitimacy. Again, biblical thinking itself carries the seeds of this idea: the relationship between God and man is more primary and more important than the relationship between man and government. In the Bible, for example, Moses lays a series of restrictions on monarchical behavior;[29] Samuel colorfully warns the Jews of the predations of kings, lamenting, "in that day you will cry out because of your king, whom you have chosen for yourselves, but the Lord will not answer you in that day."[30]

In ancient Greek and Roman thought, philosophers warred over the best governmental regime for ensuring the enforcement of virtue (that is, right reason); the ancients were far less concerned with natural rights than the imposition of natural law. Thus, Plato argued (possibly ironically) in *The Republic* for a heavy-handed regime of philosopher-kings controlling all aspects of human life, and rigidly categorizing human beings by quality;[31] Aristotle argued instead in favor of a system that would combine aspects of democracy with aristocracy, relying on Greek tradition for that model.[32] Cicero, following Aristotle's lead, championed a mixed system—a system of shared government responsibility.[33]

As Western history unfolded, it became clear that unchecked monarchies could limit rights as easily as they created new ones. If rights didn't preexist government, if they were simply whatever a government chose to enforce, then no one was safe from the government's whims.

Magna Carta, the seminal document in rethinking the balance of power between monarchs and citizens, resulted from the misgovernment of King John and the mistrust of his lords. The Thirty Years War (1618–48), which truly could be described as World War I—a war that resulted in the deaths of somewhere between four and twelve million human beings, or at least 20 percent of the population of Europe[34]—resulted in the Peace of Westphalia, in which Europeans decided to live with difference rather than compelling wholesale religious adherence. Grotius, writing contemporaneously with the Thirty Years War, suggested the dominion of rights lay in the individual, not in kings or potentates. It was not the job of governments to enforce virtue, but to enforce rights. Thus, Grotius suggested, "if a Man owes another any Thing, not in strictness of Justice but by some other Virtue, suppose Liberality, Gratitude, Compassion, or Charity, he cannot be sued in any Court of Judicature, neither can War be made upon him on that account."[35]

With the rise of the belief of individual rights came a dramatic shift in thinking about the nature of government itself. Consent became the chief element in determining the legitimacy of government, and protection of rights became the chief justification for that legitimacy. Locke stated, "the end of law is not to abolish or restrain, but to preserve and enlarge freedom: for in all the states of created beings capable of laws, where there is no law, there is no freedom."[36] Freedom, not virtue, is the goal of government; virtue is the goal of individual men, pursuing right reason—a task itself that requires freedom. And should government violate either the consent of the people or their unalienable rights, Locke stated, it would lose its mandate utterly: "whenever the legislators endeavor to take away, and destroy the property of the people, or to reduce them to slavery under arbitrary power, they put themselves into a state of war with the people, who are thereupon absolved from any farther obedience, and are left to the common refuge, which God hath provided for all men, against force and violence. . . . [Power] devolves to the people, who have a right to resume their original liberty."[37]

Boiled down into revolutionary slogans, Lockean theory became "Don't tread on me." The Declaration of Independence is a purely Lockean take on the role and power of government: to "secure these rights, Governments are instituted among Men, deriving their just powers from the consent of the governed; That whenever any Form of Government becomes destructive of these ends, it is the Right of the People to alter or to abolish it."

THE UNIONIST PHILOSOPHY OF
AMERICAN INSTITUTIONS

In order to ensure the protection of these inalienable rights and legal equality, the founders wrote a constitution. That Constitution was designed to allow for action in cases of grave national need, as demanded by broad consensus; gridlock in cases of serious disagreement, to protect minorities from the tyranny of majorities; local control on behalf of local communities, but national protection of individual rights.

The founders feared the tyranny of monarchs, but they also feared what John Adams called the "tyranny of the majority,"[38] and what James Madison termed "superior force of an interested and overbearing majority."[39] The founders believed in the people more than a self-appointed oligarchy, but they believed in individual rights more than they believed in communal self-control.

In other words, the founders correctly doubted the value of pure majoritarianism. They recognized that majorities could get the protection of rights wrong. That wasn't moral relativism—it was both realism and a recognition of objective morality, a morality that stands above the will of the people. Purely democratic government is, by definition, compulsion directed by a majority. Sometimes that's necessary. More often, it isn't. Nothing states that a democratic majority must represent the highest good. It's a competition of interests, with competing values jockeying for control of the gun. Before you grant the government the power to wield a gun, it's absolutely necessary to think about what happens if your political opponent ends up on the trigger end and you end up staring into the muzzle. The founders knew this all too well.

The founders recognized that government is not, as President Barack Obama stated, just "us."[40] They realized that we must only give the government powers enough to do what we would not mind it doing were the government controlled by those who disagree. A government gone wrong, they understood, imposing its will from above via oligarchy or below via majoritarian despotism, is the most dangerous threat to both freedom and virtue. It's just such a government that is likely to lead America down the path toward violence. If the government inserts itself into the business of guaranteeing some majority-approved "good," the minority will often refuse—correctly!—to stand for it. Or, alternatively, that minority will look—again, correctly!—to outside forces to protect its rights. What happens in a purely majoritarian state when minority rights are unfairly abused, and when the government itself is the abuser? Citizens are left with two unpleasant options: they must either leave, or they must rebel.

The founders recognized that the reason of the community might conflict with the self-evident equal rights of individuals, and that government can become a weapon of that community. That's what the American Revolution was all about: a government seeking to restrict the rights of English citizens in the name of the "good"—the good in this case being the economic health of the motherland. The rights of British citizens of America to consent via representation were simply revoked. Revolutionary firebrand James Otis led the campaign by declaring that "taxation without representation" was tyranny; his words animated a young John Adams to declare, "Every man of an immense crowded audience appeared to me to go away as I did, ready to take up arms. . . ."[41] Thomas Jefferson, in his 1775 prelude to the Declaration of Independence titled *Declaration on the Necessity of Taking Up Arms*, likewise explained, "We are reduced to the alternative of choosing an unconditional submission to the tyranny of irritable ministers, or resistance by force. The latter is our choice. We have counted the cost of this contest, and find nothing so dreadful as voluntary slavery."[42]

The invocation of slavery to describe the robbery of individual rights by a broad, overarching government regime is certainly ironic in light of the fact that many of the founders held actual slaves. But that's precisely the point: the founders trusted unity about natural rights to lead to equality sooner than attempting to force equality. Had the founders sought to abolish slavery in the Constitution, the United States never would have been formed—and the South would have become its own slaveholding country, dividing the continent *formally* between slave and free countries. The founders knew that slavery would have to be settled eventually. They consistently wrote and spoke about it. Indeed, the unwillingness of the American founders and their heirs to protect the inherent rights of slaves against their oppressive masters sowed the seeds of the Civil War; it was in the intransigence of southern majorities, insisting upon the decency of slavery, that the carnage of the Civil War was reaped.

And it was in the protection of those rights by a government *finally* fulfilling the promises of the Declaration of Independence that the Civil War finally ended slavery in America forever. Frederick Douglass, a former slave, pointed out that the glorious words of the Declaration of Independence had no application to the slave, that the celebration of the Fourth of July was a mockery to the slave: "Fellow-citizens; above your national, tumultuous joy, I hear the mournful wail of millions! whose chains, heavy and grievous yesterday, are, to-day, rendered more intolerable by the jubilee shouts that reach them." But Douglass then invoked the rights of the Declaration of Independence and the Constitution to justify his calls for liberation.[43]

Abraham Lincoln agreed. And Lincoln would carry that task to fruition—based not on the power of the majority, based not on the inherent power of government, but based on the notion that governments are instituted to protect rights, not a majority's definition of the "good."

To create a government capable of action but incapable of violation of its raison d'être—this was the goal. As Madison put it in *Federalist No. 51*, "It is of great importance in a republic not only to guard the society against the oppression of its rulers, but to guard one part of the society against the injustice of the other part. Different interests necessarily exist in different classes of citizens. If a majority be united by a common interest, the rights of the minority will be insecure."[44]

So, what did the Constitution lay forth? The practical implementation of three specific concepts: enumerated powers, checks and balances, and federalism.

The founders believed that the powers of the government had to be enumerated in order to avoid the natural consequences of necessity-based thinking. There would be times when the government had to act, the founders knew. But they also knew that governmental rationale for action tends to bleed into a broader increase in government's essential mandate. To avoid that consequence, the founders tried to lay out precisely what they believed would be the scope of governmental powers necessary to preserve the country's existence while preventing the government from becoming a Hobbesian Leviathan. To take the most obvious example, Article I of the Constitution, which explains the powers of the legislature, states, "All legislative Powers herein granted shall be vested in a Congress." This immediately implies that many powers are *not* granted. The Constitution lists only eighteen powers in Article I, Section 8, the broadest grant of power to the legislative branch.

James Madison explicitly put forward the doctrine of enumerated rights in *Federalist No. 45*: "The powers delegated by the proposed Constitution to the federal government, are few and defined. . . . [They] will be exercised principally on external objects, as war, peace, negotiation, and foreign commerce; with which last the power of taxation will, for the most part, be connected."[45] *Few and defined* is the key phrase here. He stated similarly in *Federalist No. 14*, "the general government is not to be charged with the whole power of making and administering laws. Its jurisdiction is limited to certain enumerated objects, which concern all the members of the republic, but which are not to be attained by the separate provisions of any."[46]

Many of the founders argued explicitly against the notion of a Bill of Rights, believing that such a list of rights could be read to broaden rather than to narrow governmental power—those founders understood that the tendency would be to read the Bill of Rights as a comprehensive list of things government could *not* do, with government able to act in all other areas. The structure of the Constitution, as Alexander Hamilton argued in *Federalist No. 84*, was supposed to guarantee rights in and of itself: "the Constitution is itself, in every rational sense, and to every useful purpose, A BILL OF RIGHTS."[47] Still, the antifederalists demanded a Bill of Rights that was supposed to make this fact explicit—which is why we have the Tenth Amendment: "The powers not delegated to the United States by the Constitution, nor prohibited by it to the States, are reserved to the States respectively, or to the people."

In other words: Government, stick to your job. And you have an extraordinarily specific job description.

To ensure that government would remain within its prescribed boundaries, the founders counted on two specific mechanisms: checks and balances between branches of the federal government, and federalism—state checks on federal usurpations.

Every schoolchild learns—or at least used to learn—that the three branches of the federal government are designed to prevent the other branches from superseding their designated authority. The legislature was supposed to be where laws are made; the president was supposed to execute those laws; the judiciary was supposed to interpret those laws.

Under the Constitution, the legislature was divided between a chamber represented by population (the House of Representatives) and a chamber represented by state (the Senate), creating an internal check on the most powerful branch of government. Because the Senate was originally designed so that state legislatures would appoint their senators, the Senate could represent state interests at odds with the deeper ambitions of the House. (The Seventeenth Amendment, which based election of senators on the popular vote, idiotically removed one state check on federal usurpation—now senators no longer had to fear the zealous guardians of state power in the state legislature, and could appeal directly to the population, mirroring the motives of House members, which tended toward expansion of federal power.) As Madison put it in *Federalist No. 51*, "In republican government, the legislative authority necessarily predominates. The remedy for this inconveniency is to divide the legislature into different branches; and to render them, by different modes of election and different principles of action, as little connected with each other as the nature of their common functions and their common dependence on the society will admit."[48] Furthermore, the legislature would be constrained by the veto power of the president, its own inability to execute its own laws, and a judiciary able to avoid adjudication of laws abusing the constitutional powers of the legislature.

The power of the executive was heavily circumscribed—under the Constitution, the president has only the authority to serve as commander in chief of the military, grant pardons, convene Congress in special session, receive ambassadors, and execute duly passed laws. That's it. Other powers, including treaty-making, cabinet and Supreme Court appointments, and lawmaking are all balanced out by the power of the other branches, particularly the legislature. Alexander Hamilton argued for "energy in the executive" in order to effectuate government[49]—a direct reaction to the ineffectiveness of the executive branch under the Articles of Confederation—but fully acknowledged that the presidency was not designed to be anything like a monarch, since he would be elected every four years, be subjected to oversight and impeachment, and have most of his major powers checked by the other branches.[50]

The judiciary, as the founders saw it, would be the "least dangerous branch." The founders did not want to create an oligarchic superlegislature, empowered with life appointments, to dominate the other branches of government. Hamilton observed that the judiciary "has no influence over either the sword or the purse; no direction either of the strength or of the wealth of the society; and can take no active resolution whatever. It may truly be said to have neither FORCE nor WILL, but merely judgment; and must ultimately depend upon the aid of the executive arm even for the efficacy of its judgments."[51] Hamilton reasonably posited that the judiciary *could* declare a law invalid. However, the notion that the Supreme Court represented the ultimate arbiter of constitutional meaning—that judges had been vested with the power to simply declare rights out of whole cloth or to shape the moral character of the nation via robe-garbed holy writ—is utterly baseless. Hamilton argued that should the judiciary start to become a superlegislature, it would undermine its own moral authority, and made the case for its merging into the legislature.[52] And since the judiciary's orders were not self-executing, it would take the implementation and agreement of the executive to actually effectuate judicial opinions. The *actual* algebra of judicial review was supposed to work like this: the Supreme Court would declare a law unconstitutional; the executive could either agree or disagree; the legislature could either agree or disagree. The Supreme Court had no power to cram down its opinions on the other branches. It could only act on its own behalf, just as the other branches could. This theory—departmentalism— was widespread among the founders.[53]

Finally, the Constitution was designed in the belief that local governments govern best, since they are closest to their constituents. The national government would have the ability to protect individual rights of citizens against the gravest state usurpations—this, for example, was the intent of the Article IV restrictions against states abusing the "privileges and immunities of citizens in the several states,"[54] as well as the grant of federal power to "guarantee to every state in this union a republican form of government." But the locus of government power would reside at the local level. As Madison pointed out in *Federalist No. 10*, "The smaller the society, the fewer probably will be the distinct parties and interests composing it; the fewer the distinct parties and interests, the more frequently will a majority be found of the same party."[55] Along with that homogeneity, Madison acknowledged, came the danger of oppression—which is why federalism was a useful system. In *Federalist No. 51*, therefore, Madison explained, "In the compound republic of America, the power surrendered by the people is first divided between two distinct governments, and then the portion allotted to each subdivided among distinct and separate departments. Hence a double security arises to the rights of the people. The different governments will control each other, at the same time that each will be controlled by itself."[56]

The states would not have voluntarily surrendered authority to the federal government if by doing so they felt they had written their own death warrant; nor would the citizens of the states have voluntarily submitted to a federal government unless they believed that the federal government would protect them from encroachments on their fundamental rights by the state. States were charged with day-to-day authority; the federal government was charged with overarching issues requiring vast consensus among a tremendous variety of subgroups.

Where the institutions of other countries have often preceded a philosophy, America is different: its institutions were designed in light of a founding creed. Lincoln said the Declaration of Independence was an "apple of gold," a principle that "secured our free government, and consequent prosperity"—but it required a union and a constitution, a "picture of silver, subsequently framed around it," to "adorn, and preserve it."[57] He was eminently correct: the Constitution without the Declaration is an empty vessel.

CONCLUSION

American philosophy rests on three fundamental, eternal, *unalienable* ideas.

First, American philosophy rests on the belief that human beings have real, discernable individual rights. These rights cannot be given away. They do not come from the collective. They adhere to human beings because they are part of human nature: human beings hold inestimable value, and because they have been endowed with the unique capacity to reason. Man's will is free, and his ability to choose is sacrosanct, and cannot be violated.

Second, American philosophy holds it as self-evident that all human beings are created with equal rights. This does not mean that they have equal capacities or qualities, or that their results in life will be equal, or that they begin with equal opportunities. It *does* mean that law must treat men alike, not distinguish between them on the basis of characteristics beyond their control.

Third, American philosophy demands that government protect individual rights, not override them in the name of some greater good. Rights preexist government, and thus government must be limited in orientation. It is not the job of government to usher in a reworking of mankind, but to protect mankind's nature—and to militate against conflicts between the rights of men.

These philosophical principles were put into practice by the Constitution of the United States, the most effective governmental founding document in world history. A government of enumerated powers, checks and balances, and federalism was the American solution to the problems of ineffectual government, tyranny of the majority, and invasion of individual rights.

All of these philosophical foundations were taken for granted for most of America's history. This doesn't mean that those foundations were properly fulfilled in practice, as we will explore when we review America's history. It *does* mean that Americans held in common an allegiance to the ideals, even as they were more fully realized. American philosophy was a thick cable tying us together, and allowing us to climb from the abyss of tribalism and tyranny—a cable that could be broadened to encompass more and more Americans, but a cable consistent in its tensile strength.

Until the Disintegrationists began to saw away at that cable.

CHAPTER 2

DISINTEGRATING AMERICAN PHILOSOPHY

I n January 2013, President Barack Obama stood before the Capitol
Building in Washington, D.C., and expressed his sense of what made
America unique as a nation. "What makes us exceptional," Obama
stated, "what makes us American—is our allegiance to an idea articulated in a declaration made more than two centuries ago: 'We hold
these truths to be self-evident, that all men are created equal; that they
are endowed by their Creator with certain unalienable rights; that
among these are life, liberty, and the pursuit of happiness.'"

Without any knowledge of President Obama's belief system, his invocation of the Declaration of Independence would have been well in
line with American tradition spanning back to the founding. As we've
seen, the American creed rests in the very words cited by Obama.

But President Obama's worldview stands in stark contrast to the
creed of the Declaration. The American founding philosophy rests on
natural rights to life, liberty, and property that preexist government—
rights inherent in human nature and deriving from human reason; human equality, not in attributes, but in those rights and before the law;

and government designated to protect those rights, deriving its just powers from the consent of the governed.

The Disintegrationist worldview poses a provocative and seductive alternative. The Disintegrationist worldview rejects the idea of individual rights deriving from human nature and reason. Instead, Disintegrationists suggest that human beings are innately malleable, that human nature is unfixed, and that it is therefore impossible to derive universal individual rights from a nature that simply doesn't exist. Instead, Disintegrationists posit that social changes can remake the very nature of man. And Better Humans™ can decide which social changes are necessary and important. Thus, President Obama, essentially apologizing in Hiroshima on the seventy-first anniversary of the dropping of the atomic bomb that ended World War II, suggested a "moral revolution" that could allow us to "reimagine our connection to each other as members of the human race."[1] Michelle Obama optimistically suggested in February 2008 that American souls were broken, and that her husband could fix them: "Barack is going to demand that you, too, be different." Oprah Winfrey stated that Obama would help humanity "evolve to a higher plane."[2] Obama himself— presumably one of the Better Humans™—was asked in 2004 how he would define sin. He simply stated, "Being out of alignment with my values." Not eternal values derived via the use of right reason. *His* values.[3]

Those values can be treated as the source of *true* rights—privileges dispensed from the hands of those Better Humans™, and liberty redefined as entitlements handed down by government. Thus Obama, in the same speech opening with the promise of the American founding fathers, quickly established that liberty was actually completely subjective, to be defined individually. "Being true to our founding documents," Obama stated, "does not mean we all define liberty in exactly the same way or follow the same precise path to happiness."

But, of course, that's untrue: the founders had a quite specific idea of the liberty they sought to guarantee in the Declaration of Independence and the Constitution. By reducing natural rights to an amalgam of subjective desires, Obama opened the door wide to government interventionism in the name of "rights" created by government.[4] All of which is how Obama could, for example, simply declare health care—care provided by other Americans—a "right, not a privilege."[5]

The Disintegrationist worldview further suggests that equality of rights is inapposite, since human beings are not created equal in any real respect. In the Disintegrationist view, reducing human equality to mere equality of rights would leave those worse off in a state of permanent subservience. Instead of equality before the law, Disintegrationists claim, we must aim for *inequality* before the law, so that we can achieve equality of everything else.

The Disintegrationist vision of fairness says that recognition of humanity's innate species equality—our human nature and capacity for reason—merely reinforces *true* inequality. Thus, Obama, in the same second inaugural speech in which he cited the Declaration of Independence, turned the phrase "all men are created equal" on its head, citing innate human equality as a rationale for pursuing equality of *outcome*.[6] In this, Obama was following in the footsteps of Lyndon Baines Johnson, who stated, "We seek not just freedom but opportunity. We seek not just legal equity but human ability, not just equality as a right and a theory but equality as a fact and equality as a result. . . . To this end equal opportunity is essential, but not enough, not enough."[7] In reality, of course, nobody in the United States or anywhere on earth has "the same chance to succeed," because we are all born into different circumstances, with different qualities and different hardships; true equality of opportunity would require a nightmarish invasion of rights. But the reshaping of the founding ideal of equality of rights into justification for neo-Marxist policies is both attractive and clever.

Finally, the Disintegrationist worldview scoffs at the notion of a government of limited powers, ruling by consent, and instead substitutes a government of elites, ruling for the betterment of the masses. Because there are no individual rights that preexist government, and because equality under the law is merely a pretense for injustice, the government has no "just powers"—all powers exercised by the government are just, since "the government is us," as Obama himself stated in 2013,[8] and in fact, "Government's the only thing we all belong to," as a Democratic National Convention video stated in 2012.[9] This is a far cry from James Madison's explanation that men are neither angels nor devils, and that we must be careful in allocating power thanks to that basic truth.[10]

What, then, of the founding fathers' specifically articulated beliefs in the limits of government? They were simply wrong. "What the Founders left us," Obama explained, "[is] the power, each of us, to adapt to changing times. They left us the keys to a system of self-government, the tools to do big things and important things together that we could not possibly do alone." Anything the government can do, it has the justification to do.[11] Furthermore, consent is a matter of broad and occasional voting. *Actual* policy making should be done by an army of career bureaucrats, unelected experts, and "czars" appointed to head large swaths of American life.

Not True

In short, the Disintegrationist alternative posits three specific counters to the Unionist philosophy: that human nature is infinitely malleable, and thus carries no inherent rights; that equality before the law is injustice, and that equality in every aspect of life must be our goal; and that government is the only allowable mechanism for dispensing privileges and achieving equality of outcome.

The Disintegrationist alternative does indeed represent a serious threat to the ties that bind Americans together. It reduces human beings to widgets, ready to be reshaped by the heavy hand of a top-down, unconstrained government. The Disintegrationist philosophy treats individuals as either obstacles to a broader goal or as a means to achieve that end. Either we all subject ourselves to the dictates of our betters, or we are made to heel. The seeds of reactionary revolution lie in the increasing dominance of this philosophical tyranny.

WE ARE INFINITELY MALLEABLE, AND WE CAN BE REMADE THROUGH PRIVILEGES

The American founders saw the wonder of humankind in individual reason and thus individual rights. Man's nature was to reason; man's use of reason required the rights to life, liberty, and property. Human nature could not be changed—in fact, man's unchanging nature was the source of his glory. Man was rational; therefore, his individual rights could not be trampled. Human nature provided the boundaries for human schemes. Natural law represented those boundaries.

The Disintegrationist vision sees man differently. Man is, essentially, Play-Doh. Human beings, in the Disintegrationist view, are characterized by sinless *passion* in their natural state—but we were made cruel by the institution of the very systems the founders cherished. Private property, in particular, had cramped man's capacious capacity, turning him into a chiseling Scrooge, transforming him into a cruel individualist. The only solution to this human catastrophe would be completely remaking society. And Better Humans™ were up to the task. Jean-Jacques Rousseau suggested no known limits to human nature: "We do not know what our nature permits us to be."[12] He also stated that man, being perfectible, had been destroyed by society itself, and by private property in particular: "The first man who, having enclosed a piece of ground, bethought himself of saying *This is mine*, and found people simple enough to believe him, was the real founder of civil society."[13] Similarly, in the Marxist view, human beings—aside from certain animal needs like eating and sex—were fundamentally creations of their socioeconomic structure. The truest nature of man was as a *social* animal; capitalism alienated man from others by focusing him on the acquisition of objects rather than his relations with others. Thus, communism would lead to the "return of man to himself as a *social* (i.e., human) being."[14] As Marxist psychologist Erich Fromm wrote, "For Marx the aim of socialism was the emancipation of man, and the emancipation of man was the same as his self-realization in the process of productive relatedness and oneness with man and nature. The aim of socialism was the development of the individual personality."[15]

The early-twentieth-century American progressives similarly believed in the malleability of human beings. John Dewey, perhaps the leading progressive thinker of the early twentieth century, explained that *true* liberalism—as opposed to rights-based, classical liberalism—recognized the perfectibility of human beings: "liberalism knows that an individual is nothing fixed, given ready-made. It is something achieved, and achieved not in isolation but with the aid and support of conditions, cultural and physical:—including in 'cultural,' economic, legal and political institutions as well as science and art." Shortcomings in man were therefore shortcomings in institutions.[16]

Similarly, President Woodrow Wilson, a devotee of German philosophizing and believer in the perfectibility of man by better government, suggested in a speech before the American Bar Association in 1914, "We are custodians of the spirit of righteousness, of the spirit of equal-handed justice, of the spirit of hope which believes in the perfectibility of the law with the perfectibility of human life itself."[17] Similarly, philosopher John Dewey championed the "indefinite plasticity of human nature," which had only been fossilized by "habit"— habit that could be changed. In fact, said Dewey, "If human nature is unchangeable, then there is no such thing as education and all our efforts to educate are doomed to failure. . . . The assertion that a proposed change is impossible because of the fixed constitution of human nature diverts attention from the question of whether or not a change is desirable and from the other question of how it shall be brought about."[18]

There's only one problem with this view of human nature: it is deeply, irremediably wrong. Human nature does exist; it is fixed; it is both sinful and rational. But the Disintegrationist vision is certainly attractive. It offers the beautiful absolution of personal responsibility: after all, your shortcomings can be blamed on the "system" rather than on personal choice. More important, this view of human nature offers the possibility of a utopian eschaton in which all human beings will become perfect, their hearts transformed, their identities bound up in the collective, and themselves freed from both the desires of individual inquisitiveness and the judgment of others. The Platonic Republic becomes possible, guided from above.

All we need, as a Bernie Sanders campaign organizer famously said on camera, was reeducation camps: "there is a reason Stalin had gulags."[19]

But there are, Disintegrationists recognize, opponents of this utopian vision. Those who insist human nature is indeed unchanging, and that human reason represents the heart of that nature, become enemies. They are uncaring, unfeeling, inhumane. They lack hope. They refuse to dream. They stand in the way of human happiness, which can only be found in the complete remaking of society—which, in turn, will remake humanity.

In the Disintegrationist view, to believe in boundaries on human nature is to be repressive, intolerant, constraining. Thus Disintegrationists portray recognizing baseline biological truths as bigotry, and state that unwillingness to chalk up all disparity to society flaws represents retrograde narrowmindedness. Even the processes of science that lead to conclusions at odds with Disintegrationist philosophy must be thrown out. As Harvard psychology professor Steven Pinker writes:

> The dogma that human nature does not exist, in the face of growing evidence from science and common sense that it does, has led to contempt among many scholars in the humanities for the concepts of evidence and truth. Worse, the doctrine of the blank slate often distorts science itself by making an extreme position—that culture alone determines behavior—seem moderate, and by making the moderate position—that behavior comes from an interaction of biology and culture—seem extreme.[20]

For the most radical, this means attacking the scientific process itself, robbing it of its objectivity, and deriding objectivity in science as a fanciful notion. Computational biologist Laura Boykin recently told *Wired* magazine, "Science at its core is systematically racist and sexist."[21] Donna Hughes in *Women's Studies International Forum*, for example, stated, "The scientific method is a tool for the construction and justification of dominance in the world."[22]

Timnit Gebru, research scientist at Google on ethical AI and cofounder of Black in AI, stated in the *New York Times*, "We need to change the way we educate people about science and technology. Science currently is taught as some objective view from nowhere (a term I learned about from reading feminist studies works), from no one's point of view. But there needs to be a lot more interdisciplinary work and there needs to be a rethinking of how people are taught things."[23]

In 2018, a lawsuit by wrongly fired Google employee James Damore revealed that the company had distributed a memo labeling "individual achievement," "meritocracy," "we are objective," and "colorblind racial frame" as mind-sets connected with "white dominant culture," and suggesting that managers should promote the idea that "everything is subjective."[24] Biologist Heather Heying, who was subjected to class walkouts at Evergreen State College when she taught that men are, on average, taller than women, and that "women have to be the ones that gestate and lactate,"[25] writes: "It is . . . terrifying to watch as this attempt by activist academics to dismantle logic and hypothesis, falsification and rigor, gains ground. This is conflict, plain and simple."[26] Science, as Heying writes, has been supplanted with the politics of grievance, wherein results that do not meet with the approval of the Disintegrationists are simply shouted down, and their purveyors blackballed.

How prominent are the politics of pseudoscientific grievance? In 2017, three academics set out to find the answer to that question. James Lindsay, math doctorate, Peter Boghossian, assistant professor of philosophy at Portland State University, and Helen Pluckrose, editor of AreoMagazine.com, submitted a series of twenty fake papers to peer-reviewed journals under false names. Seven of the papers were accepted by the journals; four were actually published. The papers included one on how bodybuilding was "fat-exclusionary," another suggested that dog parks represented "petri dishes for canine 'rape culture,'" and a third included a mildly rewritten section from Adolf Hitler's *Mein Kampf.* While these papers were hoaxes, other similarly ridiculous papers were not, including one 2017 paper that examined the "feminist posthumanist politics" of squirrel food. Lindsay explained, "For us, the risk of letting biased research continue to influence education, media, policy and culture is far greater than anything that will happen to us for having done this."[27] Portland State accused Boghossian of performing research on human subjects—namely, the editors of these ridiculous journals—and disciplined him.[28]

The Disintegrationist attacks on Unionist views of unchanging human nature reach their apotheosis in the Disintegrationist obliteration of biological sex. That's because the greatest and most obvious rebuttal to the Disintegrationist argument about human malleability lies in the perfectly obvious differences between men and women. These differences are not social constructs. They are not malleable. It is a simple biological fact that men are not women, and that women are not men.

Proper manners require that if you are introduced to someone as Cathy, you should call them Cathy. However, the real-world survival of mammalian species depends on this continued bifurcation between male and female, and the recognition of that bifurcation.

Yet Disintegrationists, maddened by the reality of human nature's inflexibility, see those who maintain the reality of biological sex differences as a threat.

This madness has even reached into the medical establishment. Increasingly, doctors across America are instructed to write patients' self-identified gender rather than biological sex on their medical charts. Dr. Deanna Adkins of Duke University School of Medicine claims that gender identity is "the only medically supported determinant of sex," adding, "It is counter to medical science to use chromosomes, hormones, internal reproductive organs, external genitalia, or secondary sex characteristics to override gender identity for purposes of classifying someone as male or female." This would be shocking news to doctors, who routinely use just such indicators to classify by sex—or, for that matter, the fetal testing industry, which can use DNA tests to determine prenatal sex with near-perfect accuracy.[29] Nonetheless, this rule has been applied, with dire results—such as the death of an unborn child thanks to a nurse writing "male" on a woman's chart, resulting in misdiagnosis of a pregnancy. Dr. Daphna Stroumsa of the University of Michigan wrote in the *New England Journal of Medicine* that the problem was not the mislabeling of sex. Stroumsa even stated that "he was rightly classified as a man."[30]

In matters parental, the transgender rights argument results in the belief that parents who refuse to "affirm" their child's chosen gender are actually abusers. Already, the American Medical Association has not only recommended against "conversion therapy" for children—meaning any therapy designed to make them comfortable with the gender they were born with—they outlined legislation that would ban such therapy by government diktat.[31] The next step: removing children from the homes of parents who refuse to go along with the "transition" of their child.[32]

To doubt this received wisdom—that a man can be a woman, and vice versa, all based on subjective self-identification—is to violate the precepts of decency, according to the Disintegrationists. However, they don't just think that doubt is rude. They say it's a form of violence. To write that perhaps gender dysphoria should be classified as a mental disorder is to risk your career.

Last year, for example, Brown University assistant professor of the practice of behavioral and social sciences at the School of Public Health Lisa Littman released a study. The study focused on what it described as "rapid-onset gender dysphoria": gender dysphoria that was not present in early youth, but that manifested within days or weeks in teens and young adults. According to *Science Daily*, Littman's study found that an extraordinary percentage of transgender girls were becoming transgender in tandem with members of their social group. This study, obviously, crossed taboo political lines. So Brown pulled down a press release about it, with Brown School of Public Health dean Bess Marcus issuing a letter to the entire "community." The letter explained, "The spirit of free inquiry and scholarly debate is central to academic excellence. At the same time, we believe firmly that it is also incumbent on public health researchers to listen to

multiple perspectives and to recognize and articulate the limitations of
their work. . . . The School's commitment to studying and supporting
the health and well-being of sexual and gender minority populations
is unwavering."[33]

In other words, inquiry itself was dangerous.

Human nature does exist. But the very existence of a fixed human
nature threatens the utopian visions of the Disintegrationists. If human nature is unchanging, and if it is based in reason, then individual
rights exist, as the founding fathers suggested. And if individual rights
exist, then starry-eyed visions of a remolded humanity implode. If the
individual rights stand in the way of systemic change, the individual
must give way.

It is unsurprising, therefore, that Disintegrationists deny the very
notion of individual rights, too. The founders believed that because
human nature existed, that because human nature was rooted in reason, and that because reason had to be protected, human beings were
inherently granted the rights to life, liberty, and the pursuit of happiness by Nature and Nature's God. Those rights preexisted government
and could not be invaded by government. Disintegrationists, by contrast, see such preexisting rights—often termed "negative rights," in
that they are rights *against* the government—as a chimera, or worse,
as we've seen, as an actual impediment to equality of outcome. Rights,
in the Disintegrationist view, are an obstacle to societal happiness, and
limit our greatest tool for achieving human transformation.

Again, this view has a long and inglorious history. The French
Revolutionaries held to this creed. The French Declaration of the
Rights of Man explicitly source rights in the government itself: "The
principle of all sovereignty resides essentially in the nation. No body
nor individual may exercise any authority which does not proceed directly from the nation."[34] The general will was far more important in

the shaping of individuals than individual rights, which were, after all, fictions in the absence of an enforcing regime.

Karl Marx similarly saw rights as a corrupting influence on the development of man. Individuals are by nature unequal; rights enshrine this inequality by positing that force cannot be enacted on someone against their will, despite that natural inequality. Therefore, "to avoid these defects," Marx concluded, "right, instead of being equal would have to be unequal." Rights should be overridden in favor of the collective: "from each according to his ability, to each according to his need."[35]

American progressives rallied to this banner, as well. Actual liberty, they said, required overriding negative rights. John Dewey explained that to respect negative rights would be "simply a justification of the brutalities and inequities of the existing order."[36]

Woodrow Wilson was even more explicit: he tore the Declaration of Independence to shreds. Wilson explained in 1911, "If you want to understand the real Declaration of Independence, do not repeat the preface."[37] Wilson was even more explicit in a 1907 speech: the opening sentences of the Declaration of Independence, he said, "do not afford a general theory of government to formulate policies upon. No doubt we are meant to have liberty; but each generation must form its own conception of what liberty is." The philosophy of the founders, said Wilson, was not meant to "dictate the aims and objects of any generation but their own."[38] Negative rights had, Wilson said, outlived their usefulness. This is the precise language of Barack Obama in 2013, meant to achieve the same rewriting of the bargain between citizen and government.

Carried forward, this Disintegrationist revision of God-given, inalienable *negative* rights, rights against government, into conditions created by government to supposedly *effectuate* freedom—so-called positive rights—found full flower in the language of Franklin Delano Roosevelt. In his 1944 State of the Union address, FDR explicitly rejected the sufficiency of the founding philosophy, calling negative rights "inadequate to assure us equality in the pursuit of happiness." He stated, "We have come to a clear realization of the fact that true individual freedom cannot exist without economic security and independence. 'Necessitous men are not freemen.' People who are hungry and out of a job are the stuff of which dictatorships are made." To that end, he proposed a "second Bill of Rights," including the "right to a useful and remunerative job," "the right to earn enough to provide adequate food and clothing and recreation," "the right of every family to a decent home," "the right to adequate medical care and the opportunity to achieve and enjoy good health," "the right to adequate protection from the economic fears of old age, sickness, accident, and unemployment," and "the right to a good education," among others. The American people, FDR proclaimed, had to move forward "in the implementation of these rights, to new goals of human happiness and well-being."[39]

Today, Justice Ruth Bader Ginsburg, sanctified by many on the political left as the "Notorious RBG," suggests that the United States Constitution, with its guarantees of negative rights, is actually inapposite to national flourishing. Were it up to her, she stated in 2012, she would base a new Egyptian constitution not on our Constitution, but on the "constitution of South Africa—that was a deliberate attempt to have a fundamental instrument of government that embraced basic human rights . . ."[40] Those positive rights include the rights to housing, health care, food, and social security.

None of these "rights" would have been familiar to the founders. They radically redefine the nature of rights—rights are merely goods to be provided by the government, rather than preexisting claims against the predations of government. But this redefinition has become the foundation of the Democratic Party's worldview. Senator

Bernie Sanders, Independent of Vermont, the thought-leader for the Democratic Party and a self-proclaimed democratic socialist, merely follows in FDR's footsteps when he claims that rights cannot be achieved without massive government interventionism in every sector of American life. "Today," Sanders stated in a speech at Georgetown University, "our Bill of Rights guarantees the American people a number of important constitutionally protected rights. . . . Now, we must take the next step forward and guarantee every man, woman and child in our country basic economic rights—the right to quality health care, the right to as much education as one needs to succeed in our society, the right to a good job that pays a living wage, the right to affordable housing, the right to a secure retirement, and the right to live in a clean environment."[41]

This ever-expanding rubric of pseudo-rights comes at the expense of actual rights. For example, doctors have a negative right to control their own labor—no one can force a doctor to provide service. But if we all have a positive right to health care, *someone's* negative rights will be violated: either the doctor, who will be forced to alienate his labor without his consent, or the taxpayer, who will see his own labor confiscated to pay for the doctor, or other patients, who will be deprived of their ability to obtain health care in favor of others.

Similarly, landowners have a right to build and rent apartments. But if we all have a positive right to housing, *someone's* negative rights will again be violated: either the landlord, who will be forced to rent out his apartment against his consent, or the developer, who will be forced to build more units, or the taxpayer, who will be forced to subsidize the landlord or the developer or the renter. How, in the absence of government compulsion, are these rights to be realized? Just because South Africa's constitution promises housing, food, water, and social security doesn't mean that South Africans actually *receive* these goods and services.

Once again, the Disintegrationist program is seductive. Where the founding philosophy of rights promises freedom—it promises you an adventure, but not a handout—the Disintegrationist program sets no limits on that which can be demanded from government. Furthermore, it shields its beneficiaries from claims of selfishness by hiding behind the skirts of broad-based altruism.

And, as per the usual arrangement, Disintegrationists flip the script on those who advocate for traditional Unionist rights. Those who deny a right to health care are said to be cruel and uncaring—they are portrayed as people who want to actively deprive others of the care they require. Those who suggest that goods and services cannot merely be recategorized as "rights"—that such recategorization is both intellectually dishonest and empty—are deemed heartless. To ask whether a government large enough to guarantee your wants is a government too large to be controlled is to represent pure, base cynicism. That's precisely how Barack Obama characterized his political opposition, which was supposedly too small-minded to embrace his big, systemic change: "The status quo pushes back . . . it's manufactured by the powerful and the privileged who want to keep us divided and keep us angry and keep us cynical because it helps them maintain the status quo and keep their power and keep their privilege."[42]

EQUALITY BEFORE THE LAW IS CRUELTY

Disintegrationists don't merely disagree with the proposition that human nature exists and that individual rights adhere to all of us by virtue of that nature. They disagree with the founding belief that "all men are created equal." As we have seen, the Unionist founders never meant to contend that all human beings were made equal in abilities or talents, or that they were born into similar situations. They meant that men have *species equality*—we are all equal because of our membership to the species—and that therefore, we are equal in our inalienable rights.

Disintegrationists, by contrast, suggest that all men are *not* created equal in the basic biological sense, which is true—and that equal rights, therefore, are unequal. A right to free speech, for example, does not result in equality for a man who cannot speak. A right to alienation of one's own labor, for example, does not result in equality for the man whose labor is less valuable.

Now, this argument can cut in favor of discrimination both in favor of and against various groups. In its earliest form, it cut in favor of disparate legal treatment *against* those considered unequal. Ardent advocates of slavery posited that innate human inequality provided a rationale for *disparate* treatment of human beings. John C. Calhoun, the firebrand senator, secretary of state, secretary of war, and two-time vice president, routinely disparaged the Declaration of Independence, infamously explaining on June 27, 1848, that its key proposition, that "all men are created equal," was patently false and superfluous: "[This proposition] was inserted in our Declaration of Independence without any necessity. It made no part of our justification in separating from the parent country." Calhoun described that key proposition at the heart of Americanism as a dangerous error. He stated that human beings were not free or independent, but restricted by government and socially defined. Slavery, therefore, was justified by the "debased" nature of some and the "virtue" of others. Liberty of individuals, Calhoun argued, "instead of being equal in all cases, must necessarily be very unequal among different peoples, according to their different conditions."[43]

Taking Calhoun's premise but reaching the opposite conclusion, progressive philosophers argued from the obviously unequal capabilities of men that only a radical societal restructuring could result in the desired equality and plasticity of humankind. This radical restructuring, in its ugliest form, became advocacy for eugenics. Progressive theologian Walter Rauschenbusch stated that it was the job of more advanced thinkers to "intelligently mold and guide the evolution in which we take part." Richard Ely, founder of the American Economic Association, stated that the future of civilization rested on the realization that "there are certain human beings who are absolutely unfit, and should be prevented from a continuation of their kind."[44] Ely added that science could provide "the ideal man," since the "great word is no longer natural selection but social selection." Herbert Croly, whose book *The Promise of American Life* was, according to Jonah Goldberg, the "bible of American progressivism," suggested that government ought to "interfere on behalf of the really fittest."[45]

A less hideous form of this argument against founding notions of equality has filtered down to today's politics. True equality can only be reached, in the more modern view, by a radical rethinking of our evil social system. Were our society properly drawn, all men would indeed be equal. Such is the argument of John Rawls, for example, who posited that justice lay in the "veil of ignorance." The idea is that a society is only fair if you wouldn't mind being any person in that society. This is a very appealing idea. You wouldn't want to find yourself in Nazi Germany or antebellum America without knowing which person you were in that society, so we can agree those societies were unjust—if you might just as easily be born a Jew as a brownshirt in Nazi Germany, you wouldn't be likely to defend the regime. Rawls extends this premise to countries promising negative rights, too: Would you want to be born into a country without a significant social safety net, for example, not knowing whether you might be poor, weak, or unintelligent? Wouldn't the most just society be one where all policies were aimed at equality of outcome, no matter who you were?

Rawls sought a "conception of justice that nullifies the accidents of natural endowment and the contingencies of social circumstances."[46] This form of justice would *require* unequal treatment: all "social and economic inequalities" would have to, first, result from "conditions of fair equality of opportunity" and second, "be to the greatest benefit of the least-advantaged members of society." If both prongs of this test were not fulfilled, then inequality of outcome could not be justified.[47] Politically, such Rawlsian logic has been used to justify the continuation of a heavily restricted free-market system burdened with massive redistributionism.[48]

As it turns out, it is nearly impossible to quantify or achieve "equality of opportunity" given the different characteristics and origins of individuals; endless governmental tinkering is the result, and often to dire effect. Furthermore, seeking "equality of opportunity" through disparate treatment of individuals means violating the rights of some on behalf of others. But Disintegrationists maintain that *all disparities* result from societal injustice rather than human differences. This means that using Rawlsian logic, government intervention becomes *never-ending*, since as it turns out, there will *always be disparities* between human beings. The Disintegrationist premise is simply false.

Draw a line down the middle of any room, and you will find group disparities in income, IQ, education, and age. Such disparities are not the result of societal discrimination. They are the result of statistical probability. But according to the Disintegrationists, disparities are *automatically* the result of discrimination, often relabeled under vague terms like "privilege," "institutional racism," or "patriarchalism." The Disintegrationist philosophy therefore leads to this extraordinarily destructive logic: we must have equality of opportunity, which means *unequal* rights, because people are not inherently equal; any inequality in society is proof of inequality of opportunity. No system can survive under this logic: inequality of outcome is a feature inherent to humankind. But that's precisely the point. The system must be destroyed.

Equality before the law must knuckle under. Americans *require* discrimination to wipe out human inequality. Recently, Theodore Johnson of the Brennan Center for Justice lamented the philosophy of "color-blind constitutionalism," since conservatives have utilized it to argue against policies like affirmative action. Johnson suggested that "Americans will need to decide whether creating a nation that lives up to its founding principles is best achieved by a Constitution that sees race or one that ignores it. . . . [I]gnoring race has not made the nation more equal." There is tremendous irony to the fact that John C. Calhoun would agree.[49]

The most highly publicized version of the Disintegrationist philosophy of equality comes in the form of intersectionality. Intersectionality, as originally proposed by law professor Kimberlé Crenshaw, suggested that people can be victims of discrimination on more than one level. A black woman, for example, can be discriminated against on grounds of both race and sex. The *intersection* of your membership in particular victim groups defines you. On its face, this idea is unobjectionable: of course one person can be discriminated against for a variety of motives. But intersectionality, as practiced today, goes further. It suggests that American society is structured in a hierarchy of victimhood, in which membership in a designated victim group confers automatic disadvantage, and lack of such membership automatic privilege. In fact, we don't even examine data on victimization before determining whether your membership in a particular group makes you more likely to be victimized. Instead, we simply conflate any disparity with discrimination. Disparity, in the intersectional viewpoint, automatically springs from societal discrimination; advantage always springs from the same. We must police all those who are successful who refuse to "check their privilege," and we must force them to acknowledge that privilege or charge them with complicity in the system of oppression.[50] The rules must be unequally applied so as to wipe away the vestiges of discrimination—the evidence of said discrimination being the continued and eternal presence of inequality of outcome.

This is how due process rights become secondary to concerns about *identity*. When Justice Brett Kavanaugh was nominated to the Supreme Court, one woman came forward to accuse him of attempted rape at a drinking party back during their teenage days. Not only did her story lack *any* corroborative evidence, but the witnesses she cited denied having ever attended such a party. Nonetheless, many in the media insisted that, thanks to Kavanaugh's privilege, he ought to be convicted in the court of public opinion. Even if he hadn't done anything wrong, he was awfully similar in ethnicity and income to the kinds of people who have utilized institutional power in the past to pursue evil. When Kavanaugh became righteously indignant in his own defense, this perfectly understandable emotional reaction became the subject of think pieces suggesting that Kavanaugh was merely acting out the anger of a white male being deprived of his expected due. As Erika Smith put it for McClatchy, "he just looked like an entitled, privileged white male, whining because he's unaccustomed to losing anything—much less a lifetime appointment to the nation's highest court that he always expected to get."[51]

But equal protection under the law *is* the greatest protection for those of minority identity in the first place. Jim Crow laws are often cited as the rationale for "restorative justice" programs like affirmative action, or for the proposition that equality before the law ought to be put aside to address historic grievances. But this ignores the fact that Jim Crow was an outgrowth of discriminatory thinking, not equal protection under the law—it was a fulsome *rejection* of equal protection under the law. Here is Birmingham, Alabama's city ordinance on restaurants, passed originally in 1944:

> It shall be unlawful to conduct a restaurant or other place for the serving of food in the city, at all which white and colored people are served in the same room, unless such white and colored persons are effectually separated by a solid partition extending from the floor upward to a distance of seven feet or higher, and unless a separate entrance from the street is provided for each compartment.[52]

Equal protection under the law is a far better remedy for discrimination than restorative discrimination. Equal protection works *when it is applied*. That's precisely what happened in 1960, four years before the Civil Rights Act, when four black civil rights protesters in Greensboro, North Carolina, sat down at the lunch counter at Woolworth's. Woolworth's called the police; the police did nothing, as they were law-bound to do, since the students had violated no laws. By July 1960, Woolworth's lunch counter desegregated itself, after losing $200,000.

Equal protection does work. But Disintegrationists slander as bigots all those who stand for it, all those who don't favor the abolition of equal rights in favor of equal outcome, and even those who reject the Disintegrationist premise that all inequality is a result of inequity.

That's why former Harvard president Lawrence Summers was ousted from Harvard after pointing out that disparities between men and women in hard sciences could be due in small part to the fact that men populate the furthest extremes of the bell curve in math and science test score distribution, leading to a few men always being at the very top and bottom.[53] That's why 2020 Democratic presidential candidate Pete Buttigieg found himself being labeled a "lying mother-f***er" by one highly touted black columnist for the great sin of recognizing that not all educational disparities between racial groups are the result of systemic discrimination.[54] (Buttigieg subsequently called up the columnist in order to listen to all the reasons he was, in fact, a beneficiary of white privilege for even noting that a lack of male role models has an impact on educational trajectory.) That's why those who refuse to categorize the American justice system as an extension of slavery are classified as racially ignorant; it's why those who refuse to bow before the mythical gender wage gap are met with hysterical rage. *Disparity is discrimination* has become a Disintegrationist article of faith. And heretics will be punished.

The Disintegrationist viewpoint on human inequality has real consequences. The American Astronomical Society recently suggested that Ph.D. programs stop using the Graduate Record Exam (GRE) in physics for applicants because women were underperforming. Medical schools are being encouraged to drop the MCAT, because ethnic minorities have been underperforming.[55] In New York City, students have been protesting "segregation" at schools that use non–racially biased tests because too many Asians and whites are succeeding on those tests. Segregation used to mean legally enforced discrimination based on race; now, according to the *New York Times*, students are fighting segregation by insisting on such discrimination.[56]

Perhaps the most hilarious recent example of such Disintegrationist thinking came from MSNBC host Rachel Maddow. According to NPR, Maddow arrived at Rockefeller University to give a prize to a prominent female scientist. She observed a wall covered with pictures of men. "What's up with the dude wall?" she asked. As it turns out, every "dude" on that wall had won either a Nobel Prize or a Lasker Award. This prompted the university to redesign the wall. "It just sends the message, every day when you walk by it, that science consists of old white men," said Leslie Vosshall, a neurobiologist at Rockefeller. "I think every institution needs to go out into the hallway and ask, 'What kind of message are we sending with these oil portraits and dusty old photographs?'"[57]

Those not infected with the poison of Disintegrationist philosophy might answer, "The message that vaunted scientists ought to be honored." Those infected with Disintegrationism, however, simply point to the "dude wall" as another example of the evil of the American system.

GOVERNMENTS EXIST TO REMAKE HUMANITY

Finally, Unionist philosophy held that government was established to secure our rights, "deriving their just powers from the consent of the governed." This meant, in practice, that government powers were limited by the individual rights of citizens, and that consent could only serve to uphold such a limited government. That's why the founders focused on creation of checks and balances: even a pure majority would not be enough to change the nature of citizens' relationship with government.

But, as we've seen, the Disintegrationist philosophy believes that human nature is completely malleable, and that we have no rights apart from government—and therefore, Disintegrationism holds that there are no "just powers," and pure majoritarianism can simply uproot any purported rights. Worse, since government is designed not to protect rights to provide privileges, consent becomes of secondary concern: the aristocratic disposition of government is *better* for us, since people's selfish exercise of rights must not stand in the way of the greater good. In practice, the Disintegrationist philosophy of government boils down to a government without limits, administered by bureaucrats without accountability.

This was specifically the sort of government envisioned by Auguste Comte, who theorized that rule by experts could restructure society in beneficial ways. Liberty was to be granted only to those of sufficiently developed quality: "Liberty . . . in a reasonable proportion is . . . useful to an individual and to a people who have attained a certain degree of instruction and have acquired some habits of foresight . . . [but it] is very harmful to those who have not yet fulfilled these two conditions and have the indispensable need, for themselves as much as for others, to be kept in tutelage."[58] To determine who deserved liberty and who did not, and to usher in a greater era of human development, experts were to be tasked with the scientific remaking of society.

Dewey followed this line of thought closely, as did Croly. Dewey's watchword was "pragmatism," not democracy, and he believed that the state could act as the reeducator of the masses.[59] Liberty itself could become an obstacle to the transformation of mankind's relations with each other: "the slogans of liberalism in one period can become the bulwarks of reaction." Dewey argued that liberalism ought to be "committed to the idea of historic relativity. It knows that the content of the individual and freedom change with time."[60] What's more, Dewey stated that because "effective liberty is a function of the social conditions existing at any time . . . as economic relations became dominantly controlling forces in setting the pattern of human relations, the necessity of liberty for individuals which they proclaimed will require social control of economic forces in the interest of the great mass of individuals." Individual, preexisting rights disappear. Experts will change the conditions on the ground, creating "effective liberty."[61]

Woodrow Wilson drank deeply from this intellectual well. He explained that the state ought to have no boundaries at all:

> [O]mnipotence of legislation is the first postulate of all just political theory . . . in fundamental theory socialism and democracy are almost if not quite one and the same. They both rest at bottom upon the absolute right of the community to determine its own destiny

and that of its members. Men as communities are supreme over men as individuals. Limits of wisdom and convenience to the public control there may be: limits of principle there are, upon strict analysis, none.

Democracy, Wilson stated, was bound by "no principle of its own nature to say itself nay as to the exercise of any power." There was no "just power" to which people had to consent. And consent itself might be irrelevant to the question of *true* democracy—that is, the right of the community to determine its destiny. The question, said Wilson, was one of "organization, that is to say of *administration*."[62] And the question of administration was a question of *what worked*: "All that progressives ask or desire is permission—in an era when 'development,' 'evolution,' is the scientific word—to interpret the Constitution according to the Darwinian principle; all they ask is recognition of the fact that a nation is a living thing and not a machine."[63]

The answer to that question, obviously, lay in "expertise." Unelected bureaucrats ought to control the mechanisms of government: "administration lies outside the proper sphere of *politics*. Administrative questions are not political questions." Wilson took to its logical limit the Disintegrationist rejection of both just powers and the consent of the governed: a government supposedly for the people, but not by the people.[64]

This taste for administrative government has never waned. Gradually, limits on the powers of government have dissipated; so, too, has the accountability of government to its citizens. The alternative philosophy of Disintegrationism allows citizens to abandon their responsibility in thinking about the relationship between the individual and the state; it grants citizens the ability to delegate their judgment to supposed experts, who can run their lives. Disintegrationists then flip the script: they argue that those who stand in favor of a limited government stand *against* the people. As Barack Obama constantly put it, the question of government ought not to be the scope of government, but the effectiveness of government. Finally, Disintegrationists argue that those who stand against such pragmatism are pigheaded ideologues, unwilling to accept the capacity of government to better lives. As Obama urged, "we can't take comfort in just being cynical. We all have a stake in government success—because the government is us."[65]

THE DISINTEGRATIONIST PHILOSOPHY OF AMERICAN INSTITUTIONS

If the government is us, there is no necessity for any set rules on *how* government ought to operate. The rules should—and *must*—change for the sake of "getting things done." Indeed, that is precisely the argument made by Disintegrationists, who make no commitment to any structural consistency in government. Sometimes Disintegrationists are majoritarians; sometimes they're elitists. Sometimes they're in favor of legislature supremacy, sometimes executive supremacy, sometimes judicial supremacy. Sometimes they're fans of states' authority versus the federal government; other times, they're enemies of such authority, declaring the federal government supreme in every respect.

The question isn't balancing the interests of various groups, protecting minority rights, or ensuring consensus. There is no method to this madness. The government is the source of goodness and light, and thus must never be impeded—from any angle—in doing the business of the Disintegrationists. Disintegrationists may be in the minority or they may be in the majority, but their principles rule. Institutions are merely tools for the implementation of those principles.

This lack of consistent rule-based institutions springs directly from the Disintegrationist opposition to the founding philosophy of individual rights, protected by a government instituted for that purpose. If human nature is plastic, to be shaped by social circumstance; if the goal of government ought to be equality of outcome, not merely equality of rights; if individual rights are an obstacle to that goal rather than an inherent and inalienable element of nature; and if government must be free to effect change regardless of "consent of the governed," without reference to any "just powers," then government itself must be ad hoc, pragmatist in orientation.

In practice, this means that Disintegrationists despise the constitutional order, which often requires honest men and women to recognize that their primary policy preferences may not meet with the requirements of the constitutional process. We all live by the rules—and that is the purpose of having rules. John Adams sought "a government of laws, and not of men." Disintegrationists seek a government of morals, and not of laws. This makes government God—and God cannot be constrained by the Lilliputian restraints of small-minded human beings, militating against the power of the "common good" as interpreted by the wisest among us.

Thus Disintegrationists despise the doctrine of delegated powers in government. Disintegrationists have historically stretched the boundaries of the Constitution beyond the breaking point to grant government powers it was never delegated. Often, they cite the nonoperative clauses of the Constitution—the preface speaking of promoting the general welfare, or establishing justice—in order to attempt to rewrite the actual specifics of the document. More often, Disintegrationists merely play with the elasticity of words in order to maximize the power of government. After all, there is elasticity in language—and that elasticity, like the band of an older pair of underwear, starts to weaken beyond repair over time. Thus, in *McCullough v. Maryland* (1819), the Supreme Court found that the federal government had the

power to establish a national bank, despite any constitutional dele-
gation providing as such, under the "necessary and proper clause,"
which grants the federal government the power to "make all Laws
which shall be necessary and proper for carrying into Execution" the
actual delegated powers of the government. This was in line with the
broad logic of Alexander Hamilton, who believed that the government had the power to pass any law that "might be conceived to be
conducive" in effectuating its delegated powers.[66]

That perspective, however, has little limiting principle. Historically, the "necessary and proper" clause has been combined with the
expansion of other clauses to maximize government's scope and reach.
In *Wickard v. Filburn* (1942), for example, the Supreme Court held
that the federal government had the power under the Agricultural Ad-
justment Act of 1938 to prevent a farmer from growing wheat for his
own consumption, under the constitutional power to regulate interstate commerce—even though the grain would never reach interstate
commerce.[67] It is therefore no wonder to see Disintegrationists today
arguing that the "necessary and proper" clause—the so-called "elas-
tic clause"—gives the federal government the authority to establish
federal dominance of health care, for example.[68]

The doctrine of delegate powers has become an anomaly. It barely
exists in practice. The federal government regulates nearly every area
of our lives—and the Disintegrationists cheer that fact. The new status quo has reversed the burden of proof: before, government used to
have to make an affirmative case for why it had the authority to act;
now citizens are asked to explain why they ought to have the right to
act as they see fit.

Disintegrationists have not stopped with the attempt to destroy the
notion of delegated powers. They have sought to remove the checks
and balances between the branches. The legislative branch has been
freed, obviously, to do nearly anything—Congress can regulate our

toilet fixtures and our showerheads, whether products are labeled "meat" or not, how we obtain our birth control. But the Disintegrationists have even sought to do away with any friction within the executive branch. In recent years, Democrats have taken the measures at their disposal to alleviate legislative gridlock—in 2013, for example, Senator Harry Reid, Democrat of Nevada, ended the filibuster for judicial nominees (ironically, paving the way for Republicans to ram through *their* nominees under President Trump),[69] and Senator Elizabeth Warren, Democrat of Massachusetts, pledged to end the Senate filibuster entirely.[70] But many Disintegrationists have started to call for an end to the Senate itself, which they call undemocratic and therefore fundamentally illegitimate.[71] Ending the checks and balances between the House and Senate would radically increase the amount of legislation emanating from the two chambers, since the membership would be effectively identical.

When it comes to the executive branch, Disintegrationists make no bones about their disdain for checks and balances, either. Woodrow Wilson criticized his predecessors in the White House for deliberately avoiding a breach of the Constitution due to their mewling, "conscientious scruples." He instead suggested that "[t]he President is at liberty, both in law and conscience, to be as big a man as he can. His capacity will set the limit . . . the President has the nation behind him, and Congress has not." Wilson added, "His is the vital place of action in the system, whether he accept it as such or not, and the office is the measure of the man. . . . The framers of the Constitution made in our President a more powerful, because a more isolated, king than the one they were imitating."[72] This, of course, is ridiculous—the founders made quite clear that they *didn't* want a monarch at the head of the executive branch. But according to Wilson, the president was the only figure singular enough and powerful enough to cram through radical change.

The chief tool of the executive branch would be the permanent bureaucracy—a free-floating class of government employees, the judgment of whom would often be unquestioned. These bureaucrats would become a new superlegislature, actually writing the laws, while the legislature would become a vestigial organ of government, merely delegating broad rule-making power to these supposed experts. In 1950, fewer than 10,000 pages of regulations were published in the Code of Federal Regulations; by 2018, that number surpassed 180,000 pages.[73]

Standing at the head of a quickly growing bureaucracy, the executive branch could unify legislative, judicial, and executive functions. Under both parties, the executive branch has thus grown out of control—in 1940, approximately 443,000 Americans worked for the executive branch in nonmilitary capacity; as of 2014, that number was 1,356,000.[74] In 1800, real federal outlays per capita were $16; by 1990, they were $4,760.[75] In 1929, federal net outlays represented about 3 percent of GDP; by 2018, that number was 20 percent.[76]

What of the judiciary? The Disintegrationists have effectively argued that the judiciary ought to act in some cases as a distinct law-making body, promulgating favored social policies, and in other cases as a turnstile for unconstitutional legislation. At no point have Disintegrationists believed in the original role of the judicial power—as an honest and forthright interpreter of laws. Justice Antonin Scalia believed that the job of the judiciary was to interpret laws according to their original meaning—including the Constitution—and to craft predictable rules to prevent judges from "introduc[ing] their own personal preferences into the decision" or caving to "the will of a hostile majority."[77] Barack Obama, in appointing Justice Sonia Sotomayor to the Supreme Court, praised her as a judge mixing "a common touch, a sense of compassion, and an understanding of how the world works and how ordinary people live." He had found in her the "quality of empathy, of understanding and identifying with people's hopes and struggles."[78] In this view, judges exist not to provide consistent interpretation of rules for the benefit of citizens and potential litigants, but to *do good*. Judges are deemed legislators.

And Disintegrationist judges have taken up the mantle, placing their own personal morality above that of the Constitution and the legislature. They've flipped the script, suggesting that to adhere to the Constitution means to somehow undermine American values—that the Constitution was meant to create a mandate for judges to determine the morality of the nation. Using this oligarchic logic, judges have found a rationale for a right to the killing of the unborn buried in "emanations and penumbras" in the Constitution,[79] and in the supposed "the right to define one's own concept of existence, of meaning, of the universe, and of the mystery of human life";[80] they've excavated a federal right to same-sex marriage by rendering the Constitution itself a dead letter, important only in its willingness to declare itself irrelevant ("The generations that wrote and ratified the Bill of Rights and the Fourteenth Amendment did not presume to know the extent of freedom in all of its dimensions, and so they entrusted to future generations a charter protecting the right of all persons to enjoy liberty as we learn its meaning").[81] The question here isn't the propriety of abortion or same-sex marriage as policy. The question is why the courts have suddenly become the repository of the collective wisdom. The Disintegrationists have fulfilled Hamilton's warning, undermining the fundamental legitimacy of the courts by urging them to exceed their prescribed boundaries. And when courts fail to do so, Disintegrationists have threatened to pack them or expand their constituency in order to forge a better tool out of them.

Then there's the Disintegrationist argument against federalism. Disintegrationists argue that state authority ought to be completely overwritten by federal authority—except in cases where states and localities decide to militate in favor of Disintegrationist values. So, for example, Disintegrationists cheer on sanctuary cities at the same time they lament state bills cracking down on illegal immigration. Overall, though, Disintegrationists favor the death of states at the hands of the federal government.

Citing the oppression of minorities by state governments, Disintegrationists have insisted on the continued growth of the federal government as a check against state power. This argument at least has the benefit of historical grounding: it's certainly true that advocates of states' rights did use federalist arguments to agitate on behalf of slavery and Jim Crow. Only federal intervention, first in the Civil War, then in the Reconstruction amendments, and then finally in the Civil Rights Act and Voting Rights Act, allowed for the fruition of rights for black Americans. As political scientist William Riker wrote in 1964, "If one disapproves of racism, one should disapprove of federalism."[82] "Laboratories of democracy," in Supreme Court justice Louis Brandeis's phrasing, had become torture chambers of democracy.

But this is too broad an argument. Yes, the federal government should be capable of overruling states in cases of local tyranny—but that is not an argument on behalf of federal tyranny. The founders did accept slavery as part of the founding bargain, since the alternative would have been the division of the United States itself, but they also envisioned a federal government capable of defending individual and minority rights against the predations of state governments, as we've seen. Such intervention, they believed, would be limited in duration and scope. But Disintegrationists argue that because states have failed, this means that the federal government is a superior repository of power. That is simply a lie. So long as fundamental negative rights are protected—in this context, most important, the right to leave—local governments govern best. The same federal government capable of righteously ending state segregation was also capable of horribly interning hundreds of thousands of Japanese Americans. The federal government isn't morally superior to the state governments on the basis of size alone.

In truth, the Disintegrationist desire for a dominant federal government leads to Disintegration itself: if we don't allow Texans to live like Texans and Californians to live like Californians, why should we remain part of the same country? As the federal government becomes bigger and more powerful, local communities lose control over their own lives, and must fight harder and harder to prevent domination by those who dismiss their values and concerns. The alternative to federalism isn't national unity, but dissolution.

CONCLUSION

Step one in destroying America is convincing citizens that some people will have to sacrifice their existing rights in order for others to have new ones. It is not enough, the Disintegrationists will have to say, that the historically marginalized now be treated equally by the state. The state must step in to enforce equality of outcome in every interaction.

Saying that rights are constantly expanding and open for debate is a recipe for division. Saying that society can be improved, because children can be molded into literally anything, opens the door to life-or-death fights about what they will be molded into.

Government cannot change the hearts of human beings—it cannot make them that which they are not. Simply changing material circumstances does not result in man replacing his heart of stone with a heart of flesh. And placing government guarantees of particular privileges as the center of American life requires tyrannical overlordship, either in the form of radical redistribution, or in the form of outright coercion. Americans are thus divided between those who receive and those who are forced to give.

We are not gods, and simply do not have the capacity to rectify imbalances of innate individual qualities. We have the ability to treat everyone equally under law; we have the ability to create governments to protect individual rights. But we don't have the ability to guarantee that even two kids the same age living on the same street will start from the same point; two children growing up in the same *family* don't even start from the same point. We *certainly* don't have the ability to ensure that everyone ends at the same point. Setting up all Americans as either purveyors of a hierarchical and discriminatory system or as victims of that system, we disintegrate the ties that bind Americans together.

The treatment of government as a tool of unlimited power, too, is a massive and unforgivable mistake. Questions of the common good are *entirely* political and value driven, not merely a question of pragmatic solution making—they must be subject to the consent of the people. Even consent of the people collectively cannot override individual rights, lest tyranny be the obvious result. What's more, rule from above by the "wisest and best" has rarely resulted in an expansion of human prosperity or freedom. Usually it has resulted in the opposite. The arrogation of rule-making by a self-appointed elite ends in precisely the sort of despotism against which the founding fathers rebelled.

The Disintegrationist philosophy demonstrates its explosive tendencies when it is practically implemented in terms of government. By ripping away the core components of Unionist government—delegated powers, checks and balances, and federalism—Disintegrationist government heightens the stakes of politics while reducing our common ground. It forces Americans to fight for the high ground of power, lest they be victimized by their opponents, who seek to dominate them using the mechanisms of arbitrary and unanswerable governmental coercion.

Disintegrationist philosophy tears away at the values that have historically united Americans. But that is only the beginning. Philosophy undergirds culture—the common space we share. In the United States, our Unionist philosophy led to the creation of a culture of *rights*. Next, we turn to that culture of rights—examining the nature and source of the rights we have historically cherished, and that historically united us. Then we will turn to the Disintegrationist effort to tear away that culture as well.

CHAPTER 3

THE AMERICAN CULTURE

A merica is a country of rights.

From the very beginning, we've been a people unified by our belief in rights. That is why Lincoln invoked the founders as he called for the universality of rights applied to black slaves. It's why Martin Luther King Jr. invoked the Declaration of Independence as he attempted to blast away the façade of Jim Crow. It's why nearly every successful freedom advocate in American history has used the language of rights.

We're a country of rights because, as Unionists believe, the American philosophy rests on the fundamental, self-evident truths that human nature exists, and that we have individual rights springing from that nature; that human beings are equal in their rights under the law; and that governments are instituted to protect those rights, and not infringe upon them.

We have already discussed the constitutional framework designed by the founders to guarantee the enforcement of our rights. So, if we institute governments to protect our rights, why must we have a *culture* of rights?

The simple answer: culture is more important than the legal protections codifying that culture. When culture withers, legal frameworks become no more than parchment barriers. Rights preexist government, and so do the cultures from which rights spring.[1] That's why so many of the founders originally objected to the codification of rights in the Bill of Rights: they sought to provide a framework for the defense of rights that were *culturally imbued* in the American public. To that end, the founders were quite wary of pretending they had constructed a full list of rights—to do so, they thought, could tacitly provide permission for the government to expand its power in all *other* areas. As Alexander Hamilton explained in *Federalist No. 84*, "why declare that things shall not be done which there is no power to do?"[2] Even after enumerating certain of those rights in the Bill of Rights, the founders cautiously noted in the Ninth Amendment, "The enumeration in the Constitution, of certain rights, shall not be construed to deny or disparage others retained by the people." America's culture of rights, in other words, is far broader and deeper than the legal restrictions around such rights.

So, what was the traditional American culture of rights? Or, more precisely, what culture must prevail in order for our rights to remain robust? The Unionist culture relies on four chief characteristics: first, tolerance for the rights of others, even when exercise of those rights cuts against our benefit; second, robust social institutions inculcating virtue and providing the basis for moral unity; third, stubborn-headed willingness to defend the rights of yourself and others against tyranny; and finally, a heady sense of adventure, of economic risk-taking, of willingness to value freedom over security.

RIGHTS REQUIRE THAT WE DEFEND THE RIGHTS OF OTHERS—EVEN IF WE DON'T LIKE HOW THEY USE THOSE RIGHTS

Unionist culture requires that individuals believe in rights—and most important, that they believe in the rights of others. In fact, the maintenance of a healthy culture of American rights requires that Americans believe in the broad application of rights even *beyond* legal requirements. After all, the law provides merely the final stopgap against the violation of rights; it is not the first line of defense. The first line of defense for our rights lies in our cultural defense of and belief in each other's rights.

What is the basic nature of the rights we must respect and defend? As we have seen, the founders believed that rights preexist government, and adhere to you by nature of your existence; rights represent a claim you have against the world, based on the principle that others may not harm you. You have a right to life, because others have an obligation not to kill you. You have a right to free speech, because others have an obligation not to meet speech with violence. Legal rights only exist as protections against those who violate your rights. This includes the government. These rights, as we have discussed, were typically termed "negative" rights, in that they were rights articulated against the government.

So far, so good. But there are also actions we want others to perform to which we do not have a right—actions that, in our fuzzy thinking, we often suggest we have a "right" to demand. These "rights" have been termed "positive rights." Positive rights are not rights at all. They are demands that violate someone's negative rights. So, for example, if I demand that you serve me on my terms, I am not acting on my rights. I may be morally correct that you *should* serve me on my particular terms. But I do not have a right to your services. That would violate *your* right to control your own labor—a right you *do* have against the world. I cannot demand that you say what I want you to say. That would violate your right to speak freely—a right you *do* have against everyone else.

The easy way to tell whether you do indeed have a right against me is to bring a gun to the party. If you try to kill me, and I pull a gun on you, I'm perfectly within my rights; if you try to steal my car and I pull out a gun to detain you, I'm perfectly within my rights. If, however, I stroll into your bakery, demand that you serve me a cake, and then pull a gun on you when you refuse, I have violated your rights—not the other way around.

There is one serious question about rights, however: Why should you have the *right* to do something immoral? Why should we all defend the rights of people who act in ways of which we disapprove? Why shouldn't we all agree that certain speech, for example, is beyond the pale, and then restrict it? Why shouldn't we force businesses to cater to those to whom they wish to refuse service? Why shouldn't we force those who have more to give to those who have less? If we know what's morally right, why do *rights* matter?

This question led to the rejection of individual rights by both ancient thinkers and by modern thinkers. Ancient thinkers didn't think in terms of individual rights—they thought purely in terms of duties demanded by natural law, required by the polis. Modern thinkers disparage individual rights as obstacles to the common good. Only Enlightenment thinkers of the pre-founding and founding era saw individual rights as a self-evident truth.

Why did the founding fathers, then, embrace the idea of natural rights—even the right to act immorally, so long as you did not impose on the rights of others? They did so because they realized that while social institutions may breed the values necessary to support liberty, we can never be *certain* that the right values will prevail. The founding fathers embraced epistemological uncertainty. They recognized that your definition of values may not match mine, and that the seeds of tyranny lie in using force to impose such values.

No one has ever thought his own values morally wrong, but people are frequently wrong. That's not a morally relativistic point—that's a historically accurate point. And when a monopoly on coercion is wedded with self-assurance about morality, tyranny is the most frequent result. We've become so fond of the government gun in our day and age that we somehow believe that the government ought not merely to defend rights, but also to enforce that which we—presumably, the majority—find morally right. But that's inherently dangerous. The role of government is *not* to enforce virtue. It is to protect us from the violation of our rights by others.

That Unionist system of rights does come with risks. But the risks of a rights-based system are outweighed by the risks of a system in which the government is empowered to cram down virtue on the population.

Rights aren't important simply because they're rooted in fixed human nature, and therefore in the equality of mankind. They're important because they *work* far more consistently in protecting minorities than the blunt hand of majorities. After all, as Ayn Rand correctly stated, the smallest minority of all is the individual.

Our culture of rights—and tolerance for the expression of those rights by others—is clearest in the Unionist attitude toward free speech. The founders believed that free speech rested on the fundamental rights imprinted on human nature. Reason and free will were the necessities for any government worth establishing. Benjamin Franklin placed free speech at the center of American life and American philosophy some five decades before the Constitution was written:

Freedom of speech is a principal pillar of a free government; when this support is taken away, the constitution of a free society is dissolved, and tyranny is erected on its ruins. . . . An evil magistrate intrusted with power to punish for words, would be armed with a weapon the most destructive and terrible. Under pretence of pruning off the exuberant branches, he would be apt to destroy the tree.[3]

George Washington agreed, telling his officers in 1783, "if Men are to be precluded from offering their Sentiments on a matter, which may involve the most serious and alarming consequences, that can invite the consideration of Mankind, reason is of no use to us; the freedom of Speech may be taken away, and, dumb and silent we may be led, like sheep, to the Slaughter."[4] John Adams thought the same way: "the jaws of power are always opened to devour, and her arm is always stretched out, if possible, to destroy the freedom of thinking, speaking, and writing."[5] Adams himself violated his own prescription in promulgating the Alien and Sedition Acts—but the founding generation was devoted to the importance of free thought, understanding that only trade in thoughts could purify and clarify important matters from morals to politics.

For most of American history, the notions of free speech, a free press, and freedom of thought were generally recognized as foundational. America's broad protections of speech far outstrip any such protections in Europe. In reality, the free marketplace of ideas is the key to all forward motion, to all progress. As Justice Oliver Wendell Holmes famously wrote in his dissent in *Abrams v. United States*:

[W]hen men have realized that time has upset many fighting faiths, they may come to believe even more than they believe the very foundations of their own conduct that the ultimate good desired is better reached by free trade in ideas—that the best test of truth is the power of the thought to get itself accepted in the competition of the market, and that truth is the only ground upon which their wishes safely can be carried out. That, at any rate, is the theory of our Constitution. It is an experiment, as all life is an experiment.[6]

One can argue about the application of this principle—Holmes himself suggested that some speech amounts to incitement of violence—but the principle itself does lie at the heart of the American conception of rights. This is why Americans have been so reticent to conflate speech with violence, and to ban speech. This has been a universal tendency from left to right: both William F. Buckley and the American Civil Liberties Union used to understand that the individual right to speak freely represented the core protection of American freedom.

We recognize clearly and absolutely that any government with the authority to restrict speech and thought may use that authority to quash its own opposition. The first move of authoritarian governments is to prevent the dissemination of dissent, to restrict freedom of speech, the press, and thought on grounds of danger. And it is no coincidence that periods in world history replete with repression inevitably result in stagnation. From religious domination to feudal fiefdoms to communist suppression, closed cultures remain undeveloped; it's no shock that cosmopolitan cultures—precisely the sorts of cultures now derided for their alleged "cultural appropriation"—dominate closed cultures when the two come into conflict. Bad ideas may temporarily dominate human thought, but the founders believed that free minds would eventually make men free.

Because Americans prize freedom of speech so highly, we're willing to accept and defend more rough-and-tumble in our language and in our politics. The legal standards for incitement are high; so are the legal standards for slander and libel. We would prefer more speech to less speech. And we've typically recognized that because, in the end, we share root values; we can afford to play rougher with one another. Periods of serious speech restrictions—the Wilson administration, for example, which cracked down on antiwar dissent, or the era of Joe McCarthy—are seen as dark times in American history.

RIGHTS REQUIRE A ROBUST SOCIAL FABRIC PROMOTING VIRTUE

Now, rights alone do not guarantee virtue, as we have seen—it certainly was not virtuous to bar black students from lunch counters, or to march wearing white hoods through Skokie. And indeed, a society that promoted such activity without any cultural checks would quickly collapse into disunion and failure. Unionist culture suggests that rights must be balanced by a culture that promotes virtue. Virtue lies in our nongovernment *social fabric*. The founders were deep believers that duty lies in the morality and religion taught by strong social institutions. They recognized that rights are an excellent bulwark against tyranny, but they don't provide a strong framework for decency. Rights mean that we leave each other the hell alone, which means we *can* live together. But if we hate each other, those rights won't last for long. That's why, as the founders understood, our rights were balanced by duties. Americans historically like to cite Moses's cry to Pharaoh: "Let my people go!" The founders even considered making the national seal of the United States an image of Moses presiding over the splitting of the Red Sea. But Moses followed up that cry to Pharaoh with a rationale: "That they may serve [God]." Uncle Ben told Peter Parker that with great power comes great responsibility. With great freedom comes great responsibility, too. Our constitution was truly created for a virtuous people only—a people that felt moral obligation at the root of freedom.

That moral obligation wasn't to be imposed by government; that would violate our rights. Instead, that moral obligation would suffuse our society, be taught to our children, be inculcated by our institutions. The founding fathers believed that human beings were endowed with inalienable rights—but they were only optimistic about the exercise of those rights so long as people acted like adults. The founders believed that an immoral people with rights would slide into childish libertinism, and then into the comfortable swaddling of tyranny.

To inculcate such social values, the founders relied on robust social institutions, particularly family and church. As President John Adams wrote in a letter to the Massachusetts militia, "We have no Government armed with Power capable of contending with human Passions unbridled by morality and Religion. Avarice, Ambition, Revenge or Galantry, would break the strongest Cords of our Constitution as a Whale goes through a Net. Our Constitution was made only for a moral and religious People."[7] George Washington similarly stated in his first inaugural address, "[T]he foundation of our national policy will be laid in the pure and immutable principles of private morality . . . there exists in the economy and course of nature an indissoluble union between virtue and happiness."[8] Even James Madison, who opposed the establishment of religion in Virginia by the state, did so in order to strengthen religion; religious discrimination, he felt, would alienate people from religion and lead to conflict that would undermine public comity.[9] Thomas Jefferson, the man who coined the phrase "separation of church and state," recognized that "the moral branch of religion . . . instructs us how to live well and worthily in society."[10]

The founding fathers fully recognized that without a culture of duty to support a culture of rights, both rights and duties would collapse, requiring a massive, interventionist government—precisely the kind of government they sought to escape.

Their calls were not in vain. Alexis de Tocqueville, writing some four decades later, recognized that "Americans of all ages, all conditions, all minds constantly unite." Americans were constantly engaged in social fabric-building, joining associations "religious, moral, grave, futile, very general and very particular, immense and very small; Americans use associations to give fêtes, to found seminaries, to build inns, to raise churches, to distribute books, to send missionaries to the antipodes; in this manner they create hospitals, prisons, schools. Finally, if it is a question of bringing to light a truth or developing a sentiment with the support of a great example, they associate. Everywhere that, at the head of a new undertaking, you see the government in France and a great lord in England, count on it that you will perceive an association in the United States." He also noted that should government replace that social fabric, it would endanger the "morality and intelligence of a democratic people."[11]

Alexis de Tocqueville went further. He noted "an innumerable multitude of sects in the United States. All differ in the worship one must render to the Creator, but all agree on the duties of men toward one another." And he explained, "one cannot say that in the United States religion exerts an influence on the laws or on the details of political opinions, but it directs mores, and it is in regulating the family that it works to regulate the state." Religion, said Tocqueville, provides the ultimate check against tyranny in the United States: "at the same time that the law permits the American people to do everything, religion prevents them from conceiving everything and forbids them to dare everything."[12]

Social fabric—duty to one another—was to be provided by social institutions. Government was designed to protect individuals in their rights. Rights without duties devolve into chaos. Duties without rights devolve into tyranny. Rights and duties were, in the founding

view, utterly inseparable. That balance of rights, protected by a virtuous citizenry steeped in dutiful thinking and action, was the goal of
the founding fathers.

In order to protect the social fabric, the founders instituted a guarantee of religious freedom. Even two of America's least overtly religious founders, Thomas Jefferson and James Madison, made clear
their belief that freedom of religion was a prerequisite to the establishment of any governing order. "This right is in its nature an unalienable right," Madison wrote in 1785. "It is unalienable, because
the opinions of men, depending only on the evidence contemplated by
their own minds cannot follow the dictates of other men: It is unalienable also, because what is here a right towards men, is a duty towards
the Creator." Madison went still further, declaring that "in matters of
religion, no man's right is abridged by the institution of Civil Society
and that religion is wholly exempt from its cognizance."[13] Jefferson
put it this way in the Virginia Statute for Religious Freedom (1786):

> Almighty God hath created the mind free . . . all attempts to influ-
> ence it by temporal punishment or burthens, or by civil incapacita-
> tions, tend only to beget habits of hypocrisy and meanness, and are
> a departure from the plan of the Holy author of our religion, who
> being Lord both of body and mind, yet chose not to propagate it by
> coercions on either, as was his Almighty power to do. . . . [14]

Typically, courts have attempted to read the Bill of Rights to separate two requirements regarding religion: the requirement that Congress not establish an official religion, and that Congress not prohibit
the free exercise of religion. But this provision was meant to be read in
tandem: in order to allow for the free exercise of religion, the government could not establish any particular sect. And the founders did not

believe that religion was to be relegated to the realm of thought. They fully understood that living religiously required *exercise* of religion, not merely thought or worship. A large variety of states made specific provisions exempting religious groups from generally applicable laws, for example.

Today, this is still a hotly fought topic: Just how far may government go in regulating religious practice? But the founders would surely have been appalled by the vast majority of the discussion, because as stated, *they never believed they had created a government capable of such intrusions.* The very idea of a federal government powerful enough to mandate employer health care coverage of birth control—let alone a government powerful enough to mandate that *nuns* provide such coverage—would have been astonishing and appalling to the founders.

Religious freedom is central to American culture, and it always has been.

Tocqueville's description of America as far more deeply religious than Europe—but also religiously *practical*—has remained true for centuries. Even today, Americans are far more religious than their Western European brethren. While just 18 percent of Christians across Europe pray daily, some 68 percent of Americans do; even 27 percent of Americans who describe themselves as "nothing in particular" religiously say they believe in God absolutely. According to Pew Research, "by several measures of religious commitment, religiously unaffiliated people in the US are as religious as—or in some cases even more religious than—*Christians* throughout Western European countries."[15]

Religious Americans are also more likely to create traditional family structures than their secular counterparts. According to Pew Research, the cohort least likely to be married in the United States are atheists (36 percent) and agnostics (35 percent), compared with

Mormons (66 percent) and Presbyterians (64 percent), for example.[16] Christians also have children above replacement rates (more than two children), as opposed to Americans as a whole.[17]

Religion remains central to American life, as the founders would have predicted. Religious tolerance preceded racial tolerance in the United States by centuries; the universalism of religious tolerance actually provided the impetus toward greater racial tolerance. Religious freedom provided the ultimate basis for family and community-building throughout American history. Religion provided the chief spur to duty—and that dutiful populace allowed America's culture of rights to thrive. Should our culture turn away from religious freedom, the result won't be greater freedom, but tyranny from above.

RIGHTS REQUIRE CULTURAL WILLINGNESS TO DEFEND THEM

The founders recognized that because American rights preexisted government, and the chief individual right was the right to self-preservation, a right to keep and bear arms required protection from the government. The founders were so terrified of the tyranny of centralized government via the power of a standing army that they explicitly placed a time limit on military appropriations via Congress in Article I, Section 8 of the Constitution. That's also why the founding fathers placed great stock in the militia—groups that could be mustered at the local or state level—in order to check the ambitions of an overweening national force. These militias would comprise "our sons, our brothers, our neighbors, our fellow-citizens," said Alexander Hamilton.[18] That is why Thomas Jefferson attempted to place a clause in the Virginia Constitution that would have read, "No freeman shall be debarred the use of arms."[19] And it is why the founders enshrined a right to keep and bear arms in the Constitution: to prevent tyranny. "To disarm the people," said founder George Mason, "this is the best and most effectual way to enslave them." As David Harsanyi states, "In the writings and speeches of the American Founders, the threat of disarmament was always a casus belli."[20] That's also why a huge number of Americans owned guns at the founding of the country, despite misreporting to the contrary.[21]

Was this right considered individual? It makes no sense otherwise, considering that the militia were *summoned*, a collection of individuals. "Here every private person is authorized to arm himself," said John Adams in 1770. "Arms like laws discourage and keep the invader and the plunderer in awe," said Thomas Paine. "Horrid mischief would ensue were one half the world deprived of the use of them; for while avarice and ambition have a place in the heart of man, the weak will become a prey to the strong."[22] The majority in *DC v. Heller* (2008) found that gun rights were not merely restricted to militia membership, but extended to the *rationale for militias in the first place*: self-defense.

America's gun culture was designed to allow as many law-abiding Americans to bear arms in defense of their rights as possible. Initially, because America failed to uphold the proper, universal applications of natural rights under the Declaration of Independence, the classes of people targeted for gun removal were blacks and Native Americans.[23] As the civil rights movement progressed, securing gun rights for black Americans became a key goal—and a worthwhile goal. Occidental College historian Thaddeus Russell avers, "One of the great untold stories about the civil rights movement was that it required violent resistance from blacks to be effective."[24] Historian Charles Cobb similarly contends, "Time and again, guns have proven pivotal to the African American quest for freedom."[25]

RIGHTS REQUIRE A SPIRIT OF
ADVENTURE AND RISK-TAKING

Human beings feel the necessity to control their environment; this is the root of property rights. John Locke correctly pointed out that ownership of property is merely an extension of the idea of ownership of your own labor; when we remove something from the state of nature and mix our labor with it and join something of our own to it, we thereby make that property our own.[26] The rights of property, James Madison wrote in *Federalist No. 10*, arise from the "faculties of man."[27]

The desperate human need to control our environment, however, often results in a desire to violate the rights of others. We all need security; why not obtain economic security through violating the rights of others? The temptation to protect ourselves from the vicissitudes of life by invading the economic freedom of others has been ongoing and constant throughout human history. It has also ended in much of the worst tyranny in human history. For centuries, thinkers believed that mandarins, sitting atop an economic structure, could delegate resources to greatest effect, providing economic security for all; the result was centuries of economic stagnation. That thinking extended throughout the progressive movement in Europe, which favored centralization over decentralization, and "expertise" over freedom. The apotheosis of central planning lay, of course, in the Soviet Union, where Stalin's infamous Five-Year Plans ended with the Holodomor famine in Ukraine, and in Communist China, where Mao's Great Leap Forward entailed the deaths of some 40 million people.

In reality, the field of economics relies on diverse bases of knowledge—and, more fundamentally, on property rights and rights to one's own labor—broadly construed. Adam Smith famously praised the diffusion of interests and knowledge throughout society, pointing out that diversity of desires and skill sets led to the improvement of all. Economist Friedrich Hayek correctly pointed out that the conflict between "advocates of the spontaneous extended human order created by a competitive market" and those who demand "a deliberate arrangement of human interaction by central authority based on collective command over available resources" is rooted in the ignorance of would-be central planners about how much they do not know.[28]

To see how rights in property and economics matter to societies, merely contrast the experiences of North and South Korea. The countries are divided only by an artificial political barrier; the population is ethnically identical. Yet six decades of central planning in the North has resulted in a gross domestic product (GDP) per capita of $1,214 as of 2017;[29] in South Korea, the same period has resulted in a GDP per capita of nearly $30,000.[30] North Koreans are said to be some three inches shorter than their South Korean counterparts, thanks to food shortages; North Korean life expectancy trails South Korean life expectancy by more than a decade; 97 percent of North Korea's roads are unpaved, compared with 8 percent of South Korea's.[31]

Most of us in the West now live in so-called mixed economies— free-market economies with central planning programs stacked atop them. It's easy to mistake the redistributive gains generated by central planning with the underlying prosperity to be redistributed. But that distinction is both critical and true. There is a reason that while Bernie Sanders has called Denmark his ideal socialist country, the Danish prime minister told Harvard Kennedy School of Government in 2015, "Denmark is far from a socialist planned economy. Denmark is a market economy. The Nordic model is an expanded welfare state which provides a high level of security for its citizens, but it is also a successful market economy with much freedom to pursue your dreams and live your life as you wish."[32] Denmark is annually ranked above the United States in the Heritage Foundation index of economic freedom.[33]

The precondition to economic prosperity is rights, not government cram-downs. American culture has always recognized this truth. This is why Americans have celebrated wealth creation rather than condemning it. Tocqueville noted that Americans were a "commercial people," self-starters by nature. "The inhabitant of the United States learns from birth that he must rely on himself to struggle against the evils and obstacles of life," Tocqueville observed. "[H]e has only a defiant and restive regard for social authority and he appeals to its power only when he cannot do without it." Americans were constantly inventing, constantly creating. "Boldness of enterprise," he stated, "is the foremost cause of [America's] rapid progress, its strength, and its greatness."[34] Socialist H. G. Wells, lamenting that Americans had not embraced socialism, described them as "individualists to the marrow. . . . In America the victory of private property was complete."[35]

The American spirit has always been restless; it has always been geared toward the horizon. That was true not only in terms of territorial expansion, but in terms of economic creativity. The Unionist culture is a culture of entrepreneurialism: a culture of striving and strain, of toil and hard work, a culture that celebrates success rather than seeking to punish or confiscate it. It is not a culture that recognizes that the only guarantee in America is the guarantee of an adventure.

CONCLUSION

Tocqueville summed up the American culture in one sentence: "In the United States, they associate for the goals of public security, of commerce and industry, of morality and religion. There is nothing the human will despairs of attaining by the free action of the collective power of individuals."[36] The collective power of *individuals*: individuals strong in their rights and confident in their duties, joining together when necessary to protect themselves and better the world around them.

Our culture of rights surrounds us and suffuses our conversations. It also *unites* us. The Unionist culture brings us together in recognition that an invasion of the rights of one is an invasion of the rights of all—and in realization that only a dutiful people can sustain a free country. Our culture of rights recognizes that we must respect the rights of others, even at our own expense; that we must work to build and protect the social fabric that inculcates and reinforces virtue; that we must be willing to jump to the defense of our own rights; and that we must exercise our rights in the spirit of adventure.

But now our culture of rights is under attack. It's under attack from a Disintegrationist culture that sees individual rights as a challenge to collective power, and that treats government as the only institution worth protecting. The Disintegrationist culture warns that freedom is a rough business, cruel and unsparing. The Disintegrationist culture warns that social institutions that inculcate duty are in fact merely instruments for separating us from one another—that only a unified rule from above can instill a sense of true, broad duty sufficient to maintain unity.

Benjamin Franklin warned that trading our liberty for temporary security would be a devil's bargain. It is precisely that bargain the Disintegrationists propose.

And it is to that devil's bargain we turn next.

CHAPTER 4

THE DISINTEGRATIONIST CULTURE

O ur culture of rights is under assault.

It may not feel like that at a casual glance. After all, we talk about rights incessantly—and those rights seem to grow on a constant basis. In the past few years alone, we've gained supposed rights to health care, rights to same-sex marriage, rights to housing, rights to education. We protest for rights; we chant for rights.

And yet, in reality, our rights are diminishing. Increasingly, we do not live in a free country anymore. That's because our definition of rights has shifted dramatically from the Unionist vision—the founding vision of natural rights—and toward a Progressive-era vision of rights that transforms perception of victimhood into entitlements, entitlements into rights, and natural rights into throwaways.

Our rights used to be against the encroachments of government; now those rights are obstacles to the utopian visions of our political masters.

What's more, we're queasy about the rights we supposedly love. They're ugly and uncomfortable! Rights mean that we have to tolerate those with whom we disagree, rather than forcing our views down on them; rights mean that we have to give up the desire for control in favor of a willingness to tolerate. Rights mean that sometimes the KKK marches through Skokie, that religious bakers will sometimes refuse to bake cakes to the satisfaction of Lady Gaga, that golf clubs may be closed to women, that drag queen story hour may take place at the local library.

We may be fine with some of those things, none of those things, or all of those things.

Rights mean we don't get to stop them.

The Unionist bargain suggested that our culture of rights could be upheld only under certain conditions: we would have to tolerate the rights of others, even when the exercise of those rights didn't accord with our preferences; we would have to support the existence of robust social institutions, outside the scope of government, to inculcate and reinforce virtue; we would have to stand up and be counted when rights were invaded, and pigheadedly warn an overarching government about the consequences for such invasion; and we would have to cultivate a healthy spirit of economic freedom, a taste for adventure, celebrating risk-taking rather than quashing rights to mitigate such risks.

The Disintegrationist culture chips away at each element of the Unionist culture. Disintegrationist culture doesn't mind cracking a few eggs to make a good, government-controlled omelet. To subsume the rubric of rights under the higher priority of "common good": this is the mission. We have moved from a culture that asks whether the government should have the power to intrude on our rights to a culture that asks why we should have the right to [fill in the blank]. The burden of proof, when it comes to infringement on rights, has shifted from government to the individual.

Disintegrationist culture substitutes social or governmental compulsion for tolerance, curbing rights in order to prevent their supposed "misuse," mobilizing mobs and bending institutions to its will, grinding dissenters into dust without even reverting to the power of the law—though if the Disintegrationists had their way, law would reflect their priorities. Disintegrationist culture seeks to tear away the social fabric, seeing in it the bulwark of unjust power hierarchies, and to substitute in its place top-down government as moral arbiter. Disintegrationist culture seeks to enervate the American sense of stubborn refusal to abide by top-down diktat, to remove the tacit American threat of resistance to tyranny. And Disintegrationist culture seeks to supplant the American spirit of adventure with a spirit of tired resignation, to replace the bold quest for success with a nervous and panicked attitude of rent-seeking.

Disintegrationists have already won elsewhere. Europe has already promulgated "hate speech" laws that crack down on disfavored forms of expression; now such calls issue forth in the United States. And most people in the West *like* these sorts of limitations on our rights. After all, who wants people running around spouting the n-word? Who wants restaurants refusing to serve black people, or Jews, or gays? Who likes the image of an uber-wealthy jet-setting tech bro riding his third yacht while a single mother tries to pay for dinner? What good are rights when pitted against the possible correction of such injustices?

What's the problem with all of this? The answer is obvious: once you redefine rights to mean anything you want—and replacing rights with the "common good" does just that—rights no longer have any meaning, and you lose all protection. There is no way to ban hate speech without granting the government the power to destroy free speech. You cannot mandate "economic fairness" without giving the government the power to destroy entrepreneurship. You cannot mandate racial tolerance without granting the government the power to destroy freedom of association. There are costs to rights. But there are far greater costs to making government the arbiter of right and wrong, as well as the sole repository of power.

Furthermore, restricting our culture of rights to place power in the hands of elites inherently means conflict. It means bullying and cramdowns, both socially and governmentally. It means that every election becomes a life-or-death affair in which politics becomes war by other means.

By disintegrating the culture, Disintegrationists disintegrate the country.

RIGHTS MUST BE DESTROYED TO PREVENT THEIR MISUSE

Rights are chaotic; rights are risky. In their quest to remake humanity itself, Disintegrationists fight against individual rights. In return, they offer safety. Safety from the mob. Safety *in* the mob. Safety within the prescribed boundaries set by Better Humans™.

Disintegrationists couch their disdain for dissent by claiming that they are merely defending the vulnerable. Progress and tolerance, they argue, are themselves endangered by those who violate the ever-shifting lines of political correctness. This argument was most clearly expressed in the 1960s by Professor Herbert Marcuse, who advocated that free speech be cast aside for dominant groups, since free speech benefited those groups: "Liberating tolerance, then, would mean intolerance against movements from the Right and toleration of movements from the Left . . . liberation of the Damned of the Earth presupposes suppression not only of their old but also of their new masters."[1] Today, Eve Fairbanks makes the same argument in the *Washington Post*, complaining that those who call for civil conversation are the equivalent of southern slaveholders, who she argues "stressed the importance of logic, 'facts,' 'truth,' 'science' and 'nature' much more than Northern rhetoricians did."[2] (She must have forgotten about the part where the slaveholders were holding *slaves*, and then proceeded to fight America's bloodiest war in order to preserve slavery.)

The argument has been made with the force of law in Europe, where countries, reacting to the rise of evil groups like the Nazis, have sought to curb free speech in order to curb that rise (in Germany, this argument has been termed *streitbare demokratie*—defending democracy by restricting rights via government). By characterizing opposing views as inherently discriminatory and dangerous, we can promulgate "hate speech" regulations to prevent the speech we don't like; culturally, we must impose heavy social sanctions on those who allegedly endanger the "common good." But, as writer Kevin Williamson notes:

> The alternative to a free-speech culture is neither principled mediation of political discourse by enlightened philosopher-kings nor the imposition of rational rules governing what can be said and thought and what cannot. The real-world alternative is an endless ad-hocracy dominated by tribalism and social affinity, deployed as a political weapon and subordinate to political discipline.[3]

In practical terms, we are indeed losing our free speech *culture*. This means losing both our ability to speak freely—or at least, the ability to speak in an environment of proportionality, in which outrage meets offense—and our cultural due process rights. Our careers, our livelihoods, our social groups—all can be destroyed overnight, for any reason or no reason at all. It's not enough that we have the right to free speech. We must be eager to counter free speech with free speech. As we'll see, too many Americans want less speech instead of more.

This mentality used to be foreign to political liberals in the United

States, who once suggested agreement with the phrase apocryphally attributed to Voltaire: "I disapprove of what you say, but I will defend to the death your right to say it." Not anymore. The American Civil Liberties Union once defended the right of Ku Klux Klan members to march in heavily Jewish Skokie, Illinois. But today, the ACLU has switched sides: in 2018, an internal ACLU memo explained, "Our defense of speech may have a greater or lesser harmful impact on the equality and justice work to which we are also committed . . . as an organization equally committed to free speech and equality, we should make every effort to consider the consequences of our actions, for constitutional law, for the community in which the speech will occur, and for the speaker and others whose speech might be suppressed in the future." Speech, you see, may be offensive to minority groups. And that would be oppressive. So the nation's premier free speech defense organization has decided to give up on its free speech absolutism in favor of *streitbare demokratie*.[4] (For what it's worth, the ACLU has also given up on due process: they came out against Title IX procedures that would guarantee defense rights to students accused of sexual assault and harassment, explaining that such procedures "inappropriately favor[] the accused." So much for due process.)[5]

Yes, it's not illegal to tweet something dumb, or to present an idea that doesn't fit within the increasingly narrow Overton window of acceptable discourse. The First Amendment, which concerns governmental crackdowns on free speech, isn't implicated. But our *culture* certainly is.

Now, social pressure was always a key force in American life, as it has been for all of human history. In the early nineteenth century, Tocqueville suggested that the "civilization of our age has refined the arts of despotism." In democratic republics, he wrote, "the body is left free, and the soul is enslaved." The mob, according to Tocqueville, would allow freedom of thought—but only to an extent:

He says, "You are free to think differently from me, and to retain your life, your property, and all that you possess; but if such
be your determination, you are henceforth an alien among your
people. You may retain your civil rights, but they will be useless
to you, for you will never be chosen by your fellow-citizens if you
solicit their suffrages, and they will affect to scorn you if you solicit
their esteem. You will remain among men, but you will be deprived
of the rights of mankind. Your fellow-creatures will shun you like
an impure being, and those who are most persuaded of your inno-
cence will abandon you too, lest they should be shunned in their
turn. Go in peace! I have given you your life, but it is an existence
incomparably worse than death.". . . . The majority lives in the perpetual practice of self-applause, and there are certain truths which
the Americans can only learn from strangers or from experience.[6]

In Tocqueville's view, the despotism of the mob could be just as intimidating and as damaging as despotism from tyrants. He explicitly
stated that because the people of America were "decent and orderly,"
such social sanctions were used in "judicious" fashion—he called the
"advantages derived from the exercise of this power . . . unquestion-
able." But he feared that such social pressure could be turned to nefarious ends.[7]

In this, as in much else, Tocqueville proved dispiritingly correct.
Social pressure has now been deployed on a fearsome scale. Once,
mobs required a cause to motivate them; today social media provides
a meeting place for the mob. Only later is the cause provided. Twitter resembles nothing so much as a malfunctioning immune system,
a group of white blood cells in search of a virus, reacting instead to
the wrong stimuli: we will mob *anyone*. Our culture of free speech
has an autoimmune disease. Thus, a corporate communications executive at IAC finds her life ruined for a silly joke tweet, with millions

virtue-signaling at her expense;[8] the chief executive of Mozilla Firefox finds his job ripped from him because motivated "journalists" uncovered his history of donations to pro–traditional marriage causes;[9] the kids of Covington Catholic High School find themselves under the harsh glare of the media for the crime of wearing Trump hats and *standing still* while being screamed at by Black Hebrew Israelites and confronted by a militant Native American activist.[10] We are a swarming culture, finding encouragement and social satisfaction in tearing down isolated targets. We check the trends on Twitter to determine who will be our collective target—and then we forget about them the next day.

All of this has been raised to the level of religious observance. That new woke religion comes along with the creed of political correctness, with the priesthood of media elites, with sin and absolution and sainthood defined by the mob. No longer does man labor to serve a father in heaven; now he serves to win the praise of his cultural betters. The new church persecutes heretics and rewards true believers. Those who dissent are subjected to witch trials or public confessions. Mob mentality actually provides its own ersatz social fabric—we find solidarity with one another in tearing down those who violate the ever-shifting lines of political correctness.

The mob doesn't want to discuss an unpopular view; they want to silence it. The mob doesn't want an explanation; excuses make them angrier. The mob doesn't want an apology; blood in the water encourages a feeding frenzy.

While it's certainly true that, as Tocqueville pointed out, social sanctions can be *good*—they can reinforce virtue and disincentivize vice—the social media mob is now *so* overzealous that we have moved beyond virtue and vice, into the realm of pure intimidation of *mainstream* ideas. Our social strictures were once reserved for larger issues. Now they accompany minor failings—or no failings at all, since the lines of appropriateness change randomly, seemingly without pattern or consistency. When the boundaries of the Overton window shrink in jarring fashion, without rhyme or reason, the simplest solution becomes silence.

Take the case of Kevin Williamson, for example. Williamson was hired by *The Atlantic* because he's a talented conservative writer with a hilariously misanthropic attitude. During the hiring process, Williamson told editor in chief Jeffrey Goldberg that there would be blowback from the political left for his hiring; Goldberg assured Williamson that any blowback would be weathered. Instead, after audio was "resurfaced" by Williamson's political opponents in which Williamson discussed legal punishment for abortion, *The Atlantic* reversed itself and fired Williamson. Now, Williamson's views were no secret—he's been writing about them for years. But the ease with which those who specifically work within the political oeuvre can be destroyed, made untouchable for mainstream consumption, is frightening.

The Atlantic is a private company, and if they think it will hurt their bottom line to hire a writer, that's capitalism. They obviously didn't. That's why they hired him. But a few Twitterati attacked, and *The Atlantic* caved. There were no reports of advertisers dropping *The Atlantic* over Williamson's hiring; *The Atlantic* probably received few actual subscription cancellations. The mere threat of a social media mobbing—and the upset from the interns at *The Atlantic*—drove Williamson out of the publication.

The social media mob will come for anyone and everyone—even, hilariously enough, those who have egged on the social media mob. Sarah Silverman—a wild-eyed leftist who once suggested that concern over political correctness was merely a sign of "being old"[11]— revealed just a few years later that she had lost jobs over old skits in which she put on black makeup in order to demonstrate the difficulty of being black in America. Silverman blasted "cancel culture" and "righteousness porn," explaining, "It's like, if you're not on board, if you say the wrong thing, if you had a tweet once, everyone is, like, throwing the first stone. It's so odd. It's a perversion. It's really, 'Look how righteous I am and now I'm going to press refresh all day long to see how many likes I get in my righteousness.'"[12] Silverman isn't alone. Comedians ranging from Dave Chapelle to Aziz Ansari to Bill Burr are now noting openly that our swarming culture is threatening our culture of free speech. The only comedy we can celebrate is comedy that Isn't Funny, Guys™: the scoldish stylings of Hannah Gadsby,[13] or more tellingly, the heavy political propaganda of Jimmy Kimmel, rewarded with the cover of *New York* magazine and the label, "Suddenly, He's Walter Cronkite."[14] Clap-ter has replaced laughter; applause for political agreement has supplanted comedy entirely.

Woke-scolding has become the order of the day. And it has real-world consequences, particularly when the mob targets corporations. Williamson points out that the key social institution that now builds our social fabric is our job environment. We hang out with those with whom we work. And we work at corporations. And corporations have but one object: to maximize profit. That means that corporations are both risk-averse and conformity-driven. This sort of risk-averse and conformity-driven culture runs in direct conflict with a culture of free speech: corporations are particularly susceptible to swarming tactics. Better to fire a dissenter than incur the wrath of even a small group of partisan hacks.

That becomes even truer when the partisan hacks can capture the ears of legislators, who can actively harm corporations. Businesses like Chick-fil-A can be bullied into pulling donations from Christian groups by partisan actors and media activists who accuse them of discrimination against gay Americans, without evidence; the Boy Scouts can be forced to include girls.

Motivated partisan groups, well funded by large donations, have now taken advantage of this fact to bully corporations into pulling their advertising from political media. Media Matters, an organization founded by professional smear artist David Brock and advised by Hillary Clinton, has made it its mission to target those who disagree politically for destruction. To do so, they rip clips out of context, then blast them out to connected politicos—including, during the Obama years, the Obama White House—and then work with others to astroturf public pressure against advertisers, threatening boycotts in order to quiet dissent. Those boycotts never materialize, but that's of little consequence: for corporations, bad press can be enough to tank stock or cause executive heartburn.[15]

This is an easy game to play: motivated right-wing actors on Twitter can cause Bernie Sanders to pull his congressional endorsement of longtime ally Cenk Uygur simply by resurfacing old clips of Uygur—including one in which Uygur was made to look as though he had praised David Duke (he didn't).[16] This game of gotcha has real consequences. And it makes honest discussion far more difficult.

This system of mob rule has been formalized at many of America's college campuses, where hate speech codes and "free speech zones" predominate, where microaggressions are policed and safe spaces protected. In the place we most rely on to train our citizens to counter illogic with logic, to counter opinions with facts, to counter assertions with arguments, we have decided some debates are not worth having. Our campuses, the logic goes, must be free of troublesome voices— voices that make students feel uncomfortable. That was the reason given when I was banned from speaking at DePaul University by administrators: that my mere presence might cause discomfort to some students. When I appeared at the University of California, Berkeley, protesters chanted "speech is violence"—an epigrammatic aphorism utterly unconnected from reality, but lying at the root of attempts to silence dissent.

Such idiocy isn't merely defended by college administrators; it's championed by many of them. Michael Roth, president of Wesleyan University, defends "safe spaces" as tools for inclusivity and emotional well-being, rather than tools for quashing dissent: "With mental health and suicide crises emerging on some campuses, the idea of universities taking conscious steps to protect and nurture students emotionally as well as physically should be welcome."[17] Those steps often include barring opposing viewpoints, vindictively persecuting students who cross imaginary lines, and insisting that students who do not cave to the woke mob "check their privilege." People who "invade" these safe spaces are routinely portrayed as fascists by those who spend their days ripping down posters and insisting that campuses be cleansed of minority viewpoints.

According to the Foundation for Individual Rights in Education (FIRE), about a quarter of major American campuses have "red light" policies that infringe on student free speech rights; a vast majority of institutions, nearly two-thirds, have at least a "yellow light" policy, meaning stated restrictions on speech that are overbroad or vague.[18] Such policies make students foolish and weak, incapable of either rebutting a decent argument or of dealing emotionally with the very presence of such an argument. Jonathan Haidt of New York University and Greg Lukianoff of FIRE explain, "The new protectiveness may be teaching students to think pathologically."[19]

And this means that the need for a bubble extends beyond colleges. According to a poll from the Campaign for Free Speech, 59 percent of those between the ages of 18 and 34 said that the First Amendment should be updated to curtail "hate speech" and to "reflect the cultural norms of today." More than six in ten people in that age bracket agreed that the government should have the ability to "take action against newspapers and TV stations that publish content that is biased, inflammatory, or false." More than six in ten also agreed that universities and social media should restrict speech that has the "potential to be hurtful or offensive."[20] Social media companies have taken up the call, issuing regulations on "hate speech" without defining the term, and banning a wide variety of nonapproved opinions on supremely vague grounds. In June 2019, YouTube demonetized comedian Steven Crowder after Crowder did a lisping imitation of Vox writer Carlos Maza, even though Crowder hadn't violated YouTube's terms of service.[21] In December 2019, YouTube formalized a new policy that bans videos that "maliciously insult someone" based on protected attributes. In practice, this meant that YouTube actively considered whether to take down videos that included President Trump mocking Senator Elizabeth Warren as "Pocahontas" over her false claims of Native American ancestry. This is speech-policing, pure and simple.[22]

This is only the beginning. To see the next steps for an America that embraces Disintegrationist culture, we need only look across the water. Foreign countries have taken all of this to its logical extreme: overt governmental restrictions on speech, all in the name of the emotional sensitivity of statistical rarities. In Canada, legislation has been promulgated that makes "hate speech" prosecutable—and Prime Minister Justin Trudeau wants such legislation strengthened for social media.

In Great Britain, progressive researcher Maya Forstater, a tax expert at the Center for Global Development, was fired after committing the great sin of using "offensive and exclusionary" language. What was that horrifying language? She tweeted, "men cannot change into women." That's it. She then sued under the 2020 Equality Act, stating that her verbiage was a protected philosophical belief. The Center London Employment Tribunal slapped her down, with Judge James Tayler calling her viewpoint "incompatible with human dignity and the fundamental rights of others." Tayler explained that Forstater had failed to recognize that a Gender Recognition Certificate (GRC) from the government did not merely change a person's legal sex for purposes of government documentation—it magically changed the *sex itself*. And failure to recognize the incredible, godlike power of government represented a fundamental breach of human dignity. "That is not something Ms. Forstater is entitled to ignore," the judge wrote. "Ms. Forstater's position is that even if a trans woman has a GRC, she cannot honestly describe herself as a woman. That belief is *not worthy of respect in a democratic society* [emphasis mine]." The judge continued, rejecting the fundamental freedom of expression inherent in repeating a biological truism: "Even paying due regard to the qualified right to freedom of expression, people cannot expect to be protected if their core belief involves violating others' dignity and creating an intimidating, hostile, degrading, humiliating, or offensive environment for them."[23]

It is surely a violation of human dignity to reject the most basic function of reason, logic, and truth-telling; it is surely an intimidating, degrading, humiliating, and offensive environment to be subjected to unemployment for telling the truth. But apparently the right to emotional domination of others only runs in the direction of those who insist on government-mandated fiction. Now, the CDG should be able to fire whomever it wants for whatever reason. But the judge's reasoning here makes clear that the law isn't designed to protect employers *or* employees in nonpartisan fashion. It's meant to enforce an agenda directed against certain expressions of perfectly rational speech.

This case was so outrageous that longtime liberal J. K. Rowling, author of the Harry Potter series, tweeted in Forstater's defense. Her tweet on the subject was as progressive as humanly imaginable—but still recognized the reality of biological sex: "Dress how you please. Call yourself whatever you like. Sleep with any consenting adult who'll have you. Live your best life in peace and security. But force women out of their jobs for stating that sex is real? #IStandWithMaya

#ThisIsNotADrill." This prompted an international firestorm—the social media mob activated. Rowling's left-wing bona fides did her no good: Vox, a site that declares itself an "explanatory news" outlet, headlined, "Did JK Rowling just destroy the legacy of Harry Potter with a single, transphobic tweet?" The author of this sterling explanatory journalism called her comment "transphobic BS," and added, "JKR just ruined Harry Potter, Merry Christmas."[24]

We're on the fast track toward European-style regulation of our speech. Legislation that would require completely gutting the First Amendment is gaining steam. In New York City, businesses can be fined up to $125,000 for not using the preferred pronouns of transgender people, and up to $250,000 if they do so as "a result of willful, wanton, or malicious conduct." The Equal Employment Opportunity Commission has stated that it is illegal under *federal law* to call employees by nonpreferred pronouns. As Professor Eugene Volokh of UCLA Law School says, "people can basically force us—on pain of massive legal liability—to say what they want us to say, whether or not we want to endorse the political message associated with that term, and whether or not we think it's a lie."[25]

And increasingly, prominent members of our political establishment are cheering on such regulation. In October 2019, Richard Stengel, a former editor of *Time* and Obama State Department undersecretary for public diplomacy and public affairs, openly embraced the rewriting of the First Amendment in the pages of the *Washington Post*. He explained, "Even the most sophisticated Arab diplomats that I dealt with did not understand why the First Amendment allows someone to burn a Koran. Why, they asked me, would you ever want to protect that?" Stengel declared that the First Amendment "should not protect hateful speech that can cause violence by one group against another." As per the Disintegrationists' usual case, Stengel explained that founding ideals were not based on eternal truths, but on time-bound contingencies—and that times had now changed. The First Amendment, he argued, "was engineered for a simpler era." Now, Stengel argued, we need hate speech laws. "Why shouldn't the states experiment with their own version of hate speech statutes to penalize speech that deliberately insults people based on religion, race, ethnicity and sexual orientation?" he wrote.[26]

Why? Because in a democratic republic—a republic that relies on the ability of people to consider information, to make arguments, to discuss and reflect—we cannot simply declare in top-down fashion what sort of information people are allowed to see. If we lose faith in Americans' ability to perform the most basic prerequisites of citizenship, we ought to hand over power to an aristocracy and ditch democracy altogether.

Now, our new Disintegrationist culture—a culture that frowns on free speech, and sees controversial speech as inherently risky—should at least provide unity. But clearly it's not. Do we feel *more* bound to each other, since we share in the joys of the mob? Of course not. The most obvious effect of our swarming culture—a culture focused on tearing down those who refuse to conform with the collective—is utter distrust. We don't trust our media, who are supposed to provide a check and balance against unjustified panics—they're instead egging on those panics on behalf of the Disintegrationist viewpoint. And we certainly don't trust our neighbors—they may be the ones who gain points for un-personing us on social media. We distrust our potential employees, who may see a benefit in undermining our business for the pleasure of the viewing public; our employers, who may throw us under the bus in order to avoid the displeasure of the throng; our colleagues, who may see an advantage in watching us burn. When it comes to social media—which also happens to be the chief venue for dissemination of information and ideas in the modern world—the only way to win is not to play.

The predictable result will be a renewed *militancy*, not a utopia of civility and kindness. When we prevent people from asking questions, from exploring unpopular points of view, from expressing dissent, we breed discontent. Restricting comedy, censoring speech, erasing discussion—all of these obliterate the pressure valves a civilization needs in order to cope with inevitable tensions that arise in a pluralistic democracy. The proper response to someone who disagrees is discussion; the proper response to someone who wants to silence you is a middle finger. Increasingly, thanks to our culture of silencing, we're throwing a middle finger at one another. That's what Disintegrationist culture has brought us: not unity, but rage.

SOCIAL INSTITUTIONS MUST BE TORN DOWN TO LET GOVERNMENT UNITE US

Unionist culture celebrates church and family as the bulwarks of American greatness; Disintegrationists attack precisely those institutions at the center of American life as the *barrier* to decency, rather than as the promulgators of decency in American society. A culture of rights is a culture of discrimination, according to the Disintegrationists. After all, Americans had touted individual rights while simultaneously maintaining slavery and Jim Crow; they had touted rights while ignoring the plight of the downtrodden. Traditional Unionist culture would freely acknowledge these truths, but point to the fallibility of human nature, and the curative power of social institutions. Disintegrationist culture, which held by the philosophical principle that human nature is malleable and defined by social institutions themselves, blamed precisely those social institutions for all failures. Unionist culture saw segregation as a failure of individuals to defend a rights-based system; Disintegrationist culture saw segregation as a *manifestation* of the rights-based system itself, maintained by institutions.

The obvious solution, then, was to tear down fundamental social institutions. And the easiest way to destroy those institutions was to treat them as obstacles to a greater, more fulfilling freedom. That different freedom was largely sexual in nature—the Disintegrationists held out the prospects of a sexual utopia, freed from the guilt and shame of family and religion, all consequences alleviated by government. The Disintegrationists offered the bargain of eternal adolescence, and called it rights. The only barriers to this paradise lay in hackneyed, bigoted social institutions that had constructed our incurably diseased social body.

This wasn't a new idea. Karl Marx recognized that church and family represented obstacles to societal reordering. He therefore proposed the destruction of family structure and religion as preconditions for that reordering. "The abolition of religion as the illusory happiness of the people is a demand for their true happiness," Marx suggested.[27] *The Communist Manifesto* called for the abolition of family in order to "replace home education by social," and castigated the institution of family as "bourgeois," a place of "exploitation of children by parents," and a place of degradation for women.[28]

The attack on family and church in the United States began in earnest with the second-wave feminist movement's assault on the institution of marriage. Prior to that assault, family and marriage had always been seen as a key component of civilization, an island of protection amid a sea of troubles. Young men and women aspired to marriage as the mark of adulthood; in the 1950s, the average man was married at the age of twenty-three and the average woman by the age of twenty.[29] While the fertility rate dropped rapidly between 1900 and the late 1930s, it jumped back near its turn-of-the-century highs in the 1950s, up as high as 3.8 children per woman—and even during the demographic valley, American fertility rates never dropped below replacement rates.[30] And while people did get pregnant out of wedlock, the expectation was that they would marry: in the 1950–54 period, while some 17.2 percent of all first babies were conceived out of wedlock, more than half of the pregnant women were married by the point their babies were born. By contrast, in the 1990–94 period, 52.8 percent of all first babies were conceived outside wedlock, and more than three-quarters of those babies were born to single mothers.[31]

The second-wave feminist movement radically reshifted the typical image of marriage, in which men were seen as providers and protectors, and women as homemakers. The second-wave feminists suggested that marriage had been promulgated by men as a patriarchal institution designed to chain women to the stove and the bedroom. True fulfillment, they suggested, lay in liberation from the institution of marriage more broadly. Betty Friedan argued in *The Feminine Mystique* (1963) that women were living with pangs of spiritual hunger, "a strange stirring, a sense of dissatisfaction, a yearning that women suffered in the middle of the twentieth century in the United States." Friedan, who characterized the role of married women as "the comfortable concentration camp," said that society had turned women into "walking corpses."[32] Naturally, Friedan added that religion provided a barrier to the destruction of this evil institution: "Women of orthodox Catholic or Jewish origin do not easily break through the housewife image; it is enshrined in the canons of their religion, in the assumptions of their own and their husbands' childhoods, and in their church's dogmatic definitions of marriage and motherhood."[33]

Simone de Beauvoir argued that women should, for their own good, be barred from motherhood: "No woman should be authorized to stay at home to raise her children. Society should be totally different. Women should not have that choice, precisely because if there is such a choice, too many women will make that one. It is a way of forcing women in a certain direction."[34] True fulfillment would come by liberating women from both men and their own biological drive toward child bearing and child rearing. Again, such fulfillment would require the destruction of religious institutions: "Christian ideology," she stated, "has contributed no little to the oppression of women." She called the Jewish tradition "savagely antifeminist" and maintained that there was no rationale for opposing abortion outside of religion, which engaged regularly in hypocrisy regarding the preciousness of human life.[35]

Meanwhile, the work of Alfred Kinsey suggested that Americans were, underneath their moral trappings, sexual fiends. His statistical methods weren't just rudimentary—they were completely erroneous. But Kinsey, who abhorred traditional morality, became the talk of the United States when he released his books *Sexual Behavior in the Human Male* and *Sexual Behavior in the Human Female*. He found— based on research disproportionately conducted on sex offenders— that nearly 70 percent of men had had sex with prostitutes, 17 percent of men on farms had copulated with the livestock, and 95 percent of single women had had abortions.[36] Kinsey suggested that there were but "three kinds of sexual abnormalities: abstinence, celibacy and de- layed marriage."[37] He routinely derided religion as the source of guilt and shame.[38]

This philosophy dovetailed nicely with that of influential Frank- furt School philosopher Herbert Marcuse, who called for a complete rewriting of human sexual relations—sex, he suggested, could break the chains of inequality. He explained that civilization itself could be overturned in favor of "the concept of a non-repressive civilization, based on a fundamentally different experience of being, a fundamen- tally different relation between man and nature, and fundamentally different existential relations."[39] He explained that with the human body liberated from its duties to capitalistic production, we would all be free to engage in the ultimate joys of sexual liberation: "No longer employed as instruments for retaining men in alienated performances, the barriers against absolute gratification would become elements of human freedom. . . . This *sensuous* rationality contains its own moral laws."[40] This perspective was eventually bumper-stickered into the slogan "Make love, not war," which rang across college campuses throughout the turbulent 1960s.

According to the Disintegrationists, to oppose all of this made you a prude—or worse, an authoritarian. Theodor Adorno posited in *The Authoritarian Personality* that "the religious heritage, the carry-over of old belief and the identification with certain denominations," still undergirds authoritarian tendencies, and that traditional family structures often veer into authoritarianism.[41] Adorno's colleague Wilhelm Reich posited that traditional family structures were themselves authoritarian and fascistic, "the foremost and most essential source of reproduction of every kind of reactionary thinking."[42]

Such ideas had been expressed before (Bertrand Russell, for example, decried traditional moral sexual norms in his 1929 work, *Marriage and Morals*). But those ideas had always been deemed impractical. With the rise of the Great Society under LBJ, the government suddenly made the consequences of such a moral uprooting palatable. In order to provide a safe harbor from the vicissitudes of reality, government was prepared to step in. Social institutions were sources of cruelty, repression, and guilt; government could make you free. Johnson recast the role of government as a provider of rights rather than a protector of them: "The truth is, far from crushing the individual, government at its best liberates him from the enslaving forces of his environment."[43] A culture of rights, balanced by duties inculcated by social institutions, was replaced by a culture of privileges, provided by government, replacing social institutions with handouts.

In the wake of the destruction of social institutions, the main solution to all problems became the government itself. This became a self-perpetuating cycle—because social institutions were no longer present to inculcate duties, rights were abused more and more, which necessitated more and more intrusions into rights by the government. Government dependency increased; government, in effect, became the grantor of *true* freedom. The wages of sin had once been death; now the wages of sin became a government check.

But the attack on religious institutions didn't stop there. Today the Disintegrationists have launched a vast assault on the last vestiges of church in the United States. During the 2020 presidential cycle, a variety of Democratic candidates suggested that religious Americans were merely bigots wearing crosses, covering for their brutality with a Bible. Democratic senator Cory Booker of New Jersey explained that religious people used their faith as a "justification for discrimination," and stated that as president, "I cannot allow" such behavior.[44] Pete Buttigieg made the same argument. Beto O'Rourke went even further, stating that he would remove nonprofit status from *any religious institution* that opposed same-sex marriage. "There can be no reward, no benefit, no tax break for anyone or any institution or organization in America that denies the full human rights, and the full civil rights, of everyone in America."[45]

This demonization of traditional churchgoing has become part and parcel of the Disintegrationist culture. Hollywood star Chris Pratt was raked over the coals for attending a church that allegedly—sin of sins!—stands for traditional marriage.[46] The Salvation Army has come under attack for the same moral stance.[47] Catholic schools across America have found themselves subjected to seething media hatred for refusing to employ married same-sex couples, which would violate their own precepts.[48] Entire states have effectively shut down Catholic adoption agencies for refusing to adopt children to anyone but traditionally married male-female couples.[49] The Obama administration attempted to force *nuns* to involve themselves in contraceptive care—claiming, of course, that nuns were violating the "rights" of employees to contraceptive care coverage, neglecting that to mandate such "positive rights" would inevitably invade the negative rights of the nuns.[50]

The final move by the Disintegrationists has been the reversal of the separation between church and state. That principle was originally designed to *protect religion*, not to turn the state into a weapon against it. But now Disintegrationists argue that religious people must never bring their values into the voting booth, that religion itself must be purged from the public conversation—that religion must, in fact, be made private. Hillary Clinton stated that the "advancement of the full participation of women in every aspect of their societies" requires "deep seated cultural codes, religious beliefs, and structural biases" to be "changed."[51] Barack Obama suggested that opposition to his candidacy was rooted in bitterness, rubes who "cling to guns or religion or antipathy to people who aren't like them."[52] Religion is an artifact of the past, something to be tossed away with the stale bread. And religious Americans must leave their cherished beliefs and faiths behind.

The destruction of religion and family leaves a gaping hole in the social fabric of the country. We no longer associate with one another; we used to hang out at church, go to bowling clubs with the people with whom we went to church, join the PTA with those same people. We used to find unity in widely accepted moral principles, taught in church, echoed at home. When the institutions that brought us together collapse, little is left to unify us. As conservative philosopher Yuval Levin has written, "At the root of the most significant problems America faces at home is the weakening of our core institutions—family and community, church and school, business and labor associations, civic and fraternal and political groups."[53] Government has grown consistently throughout American society over the past few decades; our social unity continues to collapse. It's no wonder that we find ourselves joining nasty Twitter mobs just to find other people to give us a sense of belonging.

AMERICANS MUST SURRENDER TO GOVERNMENT OVERREACH

The increasingly strident Disintegrationist culture has heightened conflict around our inherent defense of rights under the Constitution. The founders saw an inherent right to self-defense predating the creation of government—our first right was the right to self-preservation. That right was not surrendered to the government. America's gun culture is steeped in a sense of pioneering self-reliance—an understanding that in the end, we must have the ability to protect ourselves and our rights. We own guns so that the government cannot take away our guns.

The Disintegrationists, though, believe that gun rights are a threat, not merely in the hands of criminals, but in the hands of the general public. That is why, regardless of the correlation between gun ownership and gun violence—high in some areas, low in others—the Democratic call remains the same: to remove weapons from private individuals. Disintegrationists justify their desire to remove guns from American society by suggesting that advocates of gun rights are invading the right to life of others—that gun culture is about enshrining a "right to kill" rather than a right to self-defense. The student activists of Parkland, a Florida school terrorized by a mass shooter in 2018, gathered under the rubric "March for Our Lives," calling for "common-sense gun control"—never proposing a single policy that would have prevented the shooting at Parkland. Barack

Obama suggested that gun rights were robbing other Americans of their "right to peaceful assembly," their "right to life, liberty, and the pursuit of happiness."[54] Amnesty International contends, "The right to live free from violence, discrimination and fear has been superseded by a sense of entitlement to own a practically unlimited array of deadly weapons."[55]

The predictable result: open movement toward the full-scale invasion of gun rights. Failed 2019 Democratic presidential candidate Beto O'Rourke, the id of the Democratic Party, has been clearest about this: to the wild cheers of a Democratic audience, O'Rourke shouted, "Hell, yes, we're going to take away your AR-15, your AK-47."[56] O'Rourke would later suggest that such a policy would not require the police to remove guns from individual citizens, but could not explain how he would "take" guns without the use of force. The other Democratic candidates cheered O'Rourke's audacity. Even many Democrats who deny their intent to seize guns routinely reference Australia's mandatory gun buyback as a model.

The very questions asked by Disintegrationists betray their anti-rights viewpoint. They constantly repeat the same refrain: Why do you *need* a magazine so large, a grip of a particular color, sights of a particular design? Warren wrote in 2012, "No one needs military-grade assault weapons to hunt, and no one needs Rambo-style high capacity magazines to protect their family from intruders."[57] New York governor Andrew Cuomo averred in 2013, "It is simple. No one hunts with an assault rifle. No one needs 10 bullets to kill a deer. And too many innocent people have died already. End the madness now!"[58] Putting aside the mischaracterization of weapons use evident in these statements—in fact, the AR-15 is one of the most common weapons in home defense—the reversal of the burden of proof is astonishing. No one *needs* newspapers, strictly speaking; no one *needs* churches. We are rather attached to both. More important, we have *rights preexisting government*. Government must make an affirmative case for why it ought to be able to invade those rights—and that requires tailoring solutions to problems, a process Disintegrationists have steadfastly avoided, since they rarely have evidence to support their specific policy preferences on gun control.[59] We need not make an affirmative case as to why we *need* our rights.

Yet Disintegrationists assert that such rights ought to be chucked out the window wholesale. In March 2018, former Supreme Court justice John Paul Stevens openly advocated for the repeal of the Second Amendment. Stevens suggested that fear of a standing army, one of the central purposes of the Second Amendment, was "a relic of the 18th century"—as though Americans have no need to fear their government at all. Back in 1991, similarly, Chief Justice Warren Burger stated, "If I were writing the Bill of Rights now, there wouldn't be any such thing as the Second Amendment." Bret Stephens, from the Right, has suggested repealing the Second Amendment; so has Michael Moore from the Left.[60]

What if Americans don't wish to hand over their rights to the government that threatens them? This is where cultural intimidation is brought to bear. Those who stand up for gun rights are castigated as accessories to murder. The National Rifle Association, the Disintegrationists suggest, has "blood on its hands" for advocating gun rights, according to a bevy of politicians, including Connecticut governor Daniel Malloy.[61] Barack Obama, wiping away tears during a public address about executive orders curbing gun rights, claimed that the "gun lobby" was holding "Congress hostage." He accused Congress of cowardice, telling those who disagreed with him that they needed to "be brave enough to stand up to the gun lobby's lies."[62] The "gun lobby," of course, are merely public interest groups—supported by members of the public. But according to Obama, they are enemies of decency. "Those who defend the easy accessibility of assault weapons should meet these families and explain why that makes sense," he said.[63] Professor David DeGrazia of George Washington University expressed this crude and nasty sentiment this way: "It is time to say No to those who tolerate the killing of thousands of innocent Americans each year in the name of gun rights."[64] Nobody tolerates the killing of innocents. Many Americans believe that the best way to protect innocent Americans, both from criminals and from intrusions on their rights, is a well-armed, well-trained population of law-abiding Americans.

But those Americans are evil, according to many in the media. In 2018, immediately following the Parkland massacre, CNN held a town hall event. Supposedly the event was designed to facilitate conversation. In actuality, it was an Orwellian *Ten Minutes Hate* directed at then–NRA spokeswoman Dana Loesch and Republican senator Marco Rubio. The crowd shouted "murderer" at Loesch as she approached the stage. Rubio was approached by an understandably emotional student—who would later apologize for his comments—who said, "Senator Rubio, it's hard to look at you and not look down the barrel on an AR-15 and not look at [the shooter]." Meanwhile, Broward County sheriff Scott Israel, who would later be fired from his job after an investigation revealed top-level malfeasance in the law enforcement lead-up and response to the massacre, sat by and grinned—and even joined in the pile-on. CNN would later be given an award for excellence in journalism by the University of Southern California's Norman Lear Center.[65]

The intimidation doesn't end there. Corporations have become particularly effective targets for Disintegrationists. Walmart has repeatedly scaled back its gun sales under pressure from media organizations and outside lobbying groups; so have Kroger, L.L.Bean, Dick's Sporting Goods, and Fred Meyer.[66] Several of these outlets have seen serious declines in revenue. Delta Air Lines went so far as to announce it would discontinue discounts for NRA members.[67] For corporations, even participating in the exercise of rights is a dangerous business when the Disintegrationist mob puts them in the political crosshairs.

THE SPIRIT OF ADVENTURE AND FREEDOM MUST BE QUASHED

The culture of American rights rests, in the Unionist view, on a particular spirit of entrepreneurial adventure. Tocqueville described early Americans this way:

> At the extreme limits of the confederate states, on the boundaries of society and wilderness, stands a population of hardy adventurers who, in order to flee the poverty ready to afflict them under their fathers' roofs, have no fear of plunging into the solitudes of America and seeking a new native country there. Scarcely arrived at a place that will serve as a refuge for him, the pioneer hastily fells some trees and raises a cabin under the leaves. Nothing offers a more miserable aspect than these isolated dwellings. . . . All is primitive and savage around him, but he is so to speak the result of eighteen centuries of work and experience. He wears the clothing of the towns, he speaks their language; he knows the past, is curious about the future, argues about the present; he is a very civilized man who, for a time, submits to living in the middle of the woods, and who plunges into the wilderness of the New World with his Bible, a hatchet, and newspapers.[68]

For centuries, that spirit has not wavered. America's captains of industry, its small businesspeople, its inventors and its engineers, its explorers and its students—anyone who left home to pursue a dream, or took a risk to build a better life—all were motivated by the spirit of adventure. But in the Disintegrationist view, that precise spirit of adventure is a *danger* to a broader vision of a cohesive, government-organized society. Entrepreneurs are a threat to that society, disrupting the "fair" order of things. In fact, because entrepreneurialism requires economic freedom—and because economic freedom, in the Disintegrationist view, is the cause of inequality and brutality—that spirit must be dampened. Thus Disintegrationists see billionaires as the enemy; creators become leeches. As Democratic self-proclaimed socialist Representative Alexandria Ocasio-Cortez of New York recently put it, "No one ever makes a million dollars. You take a billion dollars. You didn't make those widgets! You sat on a couch while thousands of people were paid modern-day slave wages and, in some cases, real modern-day slavery. . . . You made that money off the backs of undocumented people." By contrast, Ocasio-Cortez explained, "the government is us."[69]

This is a common theme among Disintegrationists: the evils of the spirit of adventure. It is not those who sally forth who ought to be rewarded—it is those who get left behind. If Unionism promotes the spirit of adventure—if it urges daring and optimism—then Disintegrationism promotes the spirit of resignation and pessimism. Unionism asks, "What can I do to better my lot?" Disintegrationism asks, "Why have I not been provided for in the fashion I deserve?"

The answer to the Disintegrationist question is obvious: the system is rigged. Disintegrationism preaches that true justice can only be obtained when economic freedom—which is only cover for reinforcement of class hierarchy—is overthrown. Actual justice—FDR's "freedom from want"—can be attained only by chaining up the market, or overthrowing it entirely. Economic freedom, of the Unionist sort, is mere exploitation, pasted over with happy talk about creative destruction.

On the theoretical level, the Disintegrationist view is nonsensical. Economic freedom can only be perceived as exploitation if we redefine the word *exploitation* itself—a word normally associated with compulsion. And indeed, this is precisely what the political left does. They argue that because life circumstances require people to work at jobs they don't love, they are therefore "exploited" by their employers. Some borrow even further from Marx, suggesting that employee value ought to be gauged by labor rather than market demands—the so-called labor theory of value. In this view, to drastically simplify, worker X may be making $10 per hour, and it may take ten hours to churn out a widget—which means that we should price the widget at $100. But the "exploitative" boss charges $120 for the product, taking a 20 percent profit. This profit is "surplus value" pressed out of the system by greedy capitalists, which should in all fairness be owned by the workers. (To be fair, Marx borrowed his labor theory of value from economists who didn't seek to overthrow free markets, including Adam Smith and David Ricardo. Marx merely popularized the theory and made it the center of his theory of exploitation—a term neither Smith nor Ricardo ever would have considered with reference to the role of capitalists.)

There is only one rather large problem with this suggestion: in the real world, it is nonsense. That's not how markets work. We cannot determine the price of a product by calculating how much labor is "worth"—this gets the process exactly backward. After all, how do we know how much an hour of labor on a widget is worth *without knowing what consumers are willing to pay for the widget*? If a widget is worth $100 by the labor theory (ten hours of labor worth $10 per hour), but nobody wants to buy it, that worker's widget is *actually* worth precisely zero dollars—which means the worker's labor is worth precisely zero dollars. Consumers determine the value of a product by how much money they are willing to trade for that product. Few consumers—approaching zero—sit and calculate the work that goes into making a product before determining whether to buy it.

Furthermore, if indeed profit-driven capitalists were greedily exploiting labor, in the true sense of exploitation, we would never see wages rise—and yet we do. If profit-driven capitalists were capable of maintaining endless profit margins by squeezing profit from labor alone, products would not become more inexpensive and better quality—and yet they do.

So, what do these capitalist "exploiters" actually do? They take *all* the risk. They invest their own capital; they hire the laborers; they pay for the production and the marketing; they bear the brunt if the widgets never sell; they buy the raw materials and invest in the machinery and staff the HR department necessary to sell the widget. Often, they invent the widget itself.

Still, the labor theory of value is seductive. We think of ourselves as workers, and we like to think our work has inherent value. That's why we are so offended when an industry dies at the hands of a new industry, or a company shuts down and lays off workers thanks to lack of competitiveness. It isn't *fair*.

Looking for someone to blame, and unwilling to acknowledge that the downsides of creative destruction are outweighed on an epic scale by the upsides, Disintegrationists instead posit that the economy is "rigged" by the wealthy. This is the case Bernie Sanders routinely makes: "Even while macroeconomic numbers like GDP, the stock market and the unemployment rate are strong, millions of middle class and working people struggle to keep their heads above water, while the billionaire class consumes the lion's share of the wealth that we are collectively creating as a nation."[70] But the "billionaire class" isn't a class in any real sense—they are entrepreneurs, the vast majority of whom didn't even begin as millionaires. Furthermore, they are not "consuming" any wealth created by a collective, because wealth isn't created by a collective. It is created by employers working with employees, funded often by capitalists, to produce products for the satisfaction of individuals across the globe.

In reality, Sanders relies on capitalism to generate wealth, and calls for government to redistribute it. In his less guarded moments, he calls for nationalization of major industries and disparages the profit motive. This strategy has produced nothing but misery throughout human history. But at least it's "fair."

Unfortunately, broad feelings about "fairness" have nothing to do with reality. Nor do they have to do with prosperity. Free markets not only do not exploit workers generally, but they have lifted the vast majority of the globe from crippling poverty over the past four decades, provided the source of nearly every major technological innovation in history, and made nearly every product better, cheaper, and more available. Free markets rely on the wisdom of individuals, aggregated by the fact of social living. Those who attempt to crack down on that natural system leave misery and poverty in their wake.

Because of the exponential growth in prosperity since the turn of the nineteenth century due to free markets, the political left can't argue convincingly that free markets have made people worse off. Instead, they focus on income inequality, arguing that the rich are growing richer while the poor are growing . . . richer, but less quickly. Sanders simply lies about this—he argues that the poor are actually growing poorer, or at least strongly implies it—but that's simply nonsense. The truth is that those who earn lots of money don't constitute a solid class: people move in and out of income quintiles over time. Furthermore, the death of the middle class has been greatly exaggerated: the vast majority of Americans who exited the $50,000 to $100,000 income group moved *up*, into the $100,000+ group. And as for charges that the American economy is somehow less competitive than its social democratic compatriots in Europe, the U.S. economy has grown multiple times faster than economies like those of Germany, France, and Japan over the past few decades.[71]

Those who rail about the 1 percent more honestly suggest that the mere fact of economic inequality threatens the stability of society at large. Thomas Piketty, the beloved French author of *Capital in the Twenty-First Century*, suggests that free markets "are potentially threatening to democratic societies and to the values of social justice on which they are based." Yet no single economic system has generated more wealth than economic freedom, particularly for those at the *bottom end* of the economic scale. It may very well be true that the billionaire buying his sixth private jet is gaining money faster by percentage than the poverty-stricken person who just received a pay raise. But the pay raise for the person at the bottom of the scale is more important from an ethical perspective—and any system that undermines the possibility of such gains for the poor cannot be said to be socially just in any real way. Deirdre McCloskey correctly observes:

In relative terms, the poorest people in the developed economies and billions in the poor countries have been the biggest beneficiaries. The rich became richer, true. But the poor have gas heating, cars, smallpox vaccinations, indoor plumbing, cheap travel, rights for women, low child mortality, adequate nutrition, taller bodies, doubled life expectancy, schooling for their kids, newspapers, a vote, a shot at university and respect.[72]

In the end, most on the Left are forced to acknowledge the reality: markets work. So instead, they undermine markets by suggesting that capitalism is theft (Elizabeth Warren), or criticizing America's capitalist economy as "imperfect, unfair, unjust, and racist" (Beto O'Rourke).[73] It's easy to avoid the consequences of undermining the greatest engine for human prosperity in history if you pay lip service to that engine, then declare that it requires fundamental change. The subtext is obvious: free markets are evil. Hillary Clinton admitted as much in the aftermath of the 2016 election, stating that allegiance to capitalism hurts electability with the Democratic Party base.[74] And it's perfectly clear that many Democrats rally around this sentiment: in the middle of the 2020 election cycle, Elizabeth Warren began selling "Billionaire Tears" mugs, and Pete Buttigieg—a supposed moderate—trotted out his relative poverty compared to Warren's as a *credential* for the presidency, prompting comedian Trevor Noah to wryly observe, "You realize for Progressive white people [being called a wealthy person is] like being called the n-word. They're like, 'How dare you call me wealthy, I'm living comfortably, okay?'"[75]

THE BIPARTISAN ATTACK ON THE CULTURE OF RIGHTS

The Disintegrationist attack on both the institutions undergirding our culture of rights, as well as on our rights themselves, has bred a reaction. That reaction isn't libertarian in orientation. Instead, on the Right, a new movement has gained credence, arguing that rights must take a backseat to higher priorities, that our culture of rights must be discarded in favor of a culture of coercion—but coercion toward a rebuilt social fabric.

Many on the Right argue that reliance on bigger government isn't a choice, but a necessity. The "common good" requires it. They argue that the failure of social institutions hasn't been a by-product of governmental intervention, but a by-product of liberalism itself. Individual rights, in this view, fail to account for humanity's need for community properly, and thus for communal standards—individual rights turn us into atomized fragments, destined to conflict with one another. Individual rights are inherently relativistic, recognizing as they do different and competing versions of "the good."

Patrick Deneen, the author of *Why Liberalism Failed*, makes this argument most eloquently. According to Deneen, the American founders erred in "not foreseeing that their atomistic philosophy would act as a solvent on our civic institutions"; he also casts heavy aspersions at capitalism, stating, "The economic system that simultaneously is both liberalism's handmaiden and also its engine, like a Frankenstein monster, takes on a life of its own, and its processes and logic can no longer be controlled by people purportedly enjoying the greatest freedom in history."[76] Deneen argues that liberty-oriented individualists ended up making common cause with leftist thinkers pushing against traditional social mores—and that's how we ended up with an allegedly heartless capitalism alongside a sexual politics focused chiefly on individual choice rather than on responsible standards.

Government, argue these right-wing thinkers, can be repurposed as a tool to overthrow the culture of rights and rebuild failing social institutions. Terry Schilling, executive director of the American Principles Project, argued recently in *First Things*, "a new conservatism is being born—one less interested in managing our nation's decline than in using political power to promote virtue, public morality, and the common good. Conservatives need to overcome their fear of governing the nation that elected them."[77]

The crossover between Disintegrationists and "common good conservatives" becomes most obvious in the realm of economics. "Common good conservatives" argue that capitalism ought to be chained to the priorities of social conservatives. Economist Joseph Schumpeter theorized that capitalism had the tendency to break down group identity and religiously rooted social fabric in favor of individualization; he worried that such a tendency could end with citizens turning on "private property and the whole scheme of bourgeois values."[78] Many on the right nod along with the critique—and suggest that free markets must therefore be the problem. Instead of recognizing that a spiritual lack must be healed through spiritual means, these conservatives suggest that markets should be heavily restricted in order to rectify that spiritual lack, thus promoting the "common good." At *First Things*, a coalition of thinkers including Deneen and Rod Dreher of *The American Conservative* tore into free markets as tools of atomization, soul-deadening and materialism focused. In a piece titled "Against the Dead Consensus," they wrote:

> We oppose the soulless society of individual affluence. Our society must not prioritize the needs of the childless, the healthy, and the intellectually competitive. . . . We want a country that works for workers. The Republican Party has for too long held investors and "job creators" above workers and citizens, dismissing vast swaths of Americans as takers unworthy of its time. . . . [We believe in] the potential of a political movement that heeds the cries of the working class as much as the demands of capital. Americans take more pride in their identity as workers than their identity as consumers. Economic and welfare policy should prioritize work over consumption.[79]

This is language with which socialists are all too familiar, and all too comfortable; it's a basic restatement of the labor theory of value, a belief that the price mechanism should be based not on the aggregate demand represented by individual consumer preference, but on some unspecified metric of work input. Instead of allowing prices to be set by consumers, this vision of a "fairer" economy places an elite estimation of the value of work at the center of economic policy. This is the central fallacy of collectivist government. We can certainly soften the blows of Schumpeter's creative destruction. But to coercively reorient markets toward top-down estimated labor value rather than diffusely sourced market value is to overthrow them entirely. And the perspective of "common good conservatives" presents no better limiting principle than the perspective of overt Disintegrationists.

Surprisingly, Republican senator Marco Rubio of Florida, a former Tea Partier, has become a leading advocate of the "common-good capitalism" movement in economics. Such language is self-contradictory—capitalism is about individuals making decisions about their own creativity and labor, and the benefits to all that flow therefrom. We don't need the modifier "common good" to describe capitalism, which is a system that prizes consensual and mutually beneficial exchange.

But according to Rubio, economic freedom leaves Americans behind—and we must restrict and harness that freedom in order to help discrete subgroups of the American population. Rubio quoted Pope Leo XIII's encyclical *Rerum Novarum* to lay forth the principle that businesses "have an obligation to reinvest those profits productively for the benefit of the workers and the society that made it possible." But that reinvestment is typically just called investment: in new technologies, products, wages; in the stock market; in savings and loans. Rubio, it turns out, didn't mean anything as prosaic as all of that. Instead, Rubio stated that the free market had failed because it didn't provide "dignified work" or serve the "broader national interest"; he lamented the "victims of an economic re-ordering." Rubio laid the death of family at the hands of capitalism, explaining, "when an economy stops providing dignified work for millions of people, families and communities begin to erode." He lamented the creation of what he termed "pockets of prosperity" amid a more general disintegration. And he adopted the language of the political left, stating that Americans were "angry at a system that has been rigged against them by the very people who created these problems," and urging acceptance of "the indivisible tie between culture and economics."[80]

This is largely wrongheaded. There are solid reasons for cracking down on Chinese trade abuse, and interesting arguments in favor of particular sorts of tax cuts. But in the end, the collapse of American culture had nearly *nothing* to do with the sort of economy Rubio laments. America's traditional values collapsed in the 1960s, when the American economy was booming and when the sort of "dignified work" to which Rubio seems to allude—the proverbial line job at Ford—was indeed available. Even more, government involvement in the economy worsened precisely the societal breakdown Rubio decries.

Tucker Carlson speaks in more emotionally charged—and more blunt—language. Here was Carlson's explanation of capitalism in January 2019:

> Market capitalism is a tool, like a staple gun or a toaster. You'd have to be a fool to worship it. Our system was created by human beings for the benefit of human beings. We do not exist to serve markets. Just the opposite. Any economic system that weakens and destroys families is not worth having. A system like that is the enemy of a healthy society. . . . We are ruled by mercenaries who feel no long-term obligation to the people they rule.[81]

This is patently erroneous on nearly every level. We cannot "make the market work for us" any more than we can "make free speech work for us." Both already work for us, because neither invades our rights; the greatest guarantee that free speech, like free markets, is an asset is to live virtuously and urge others to do so as well.

Carlson's key error—the belief that there is a collective "we" damaged by the free market—is obvious. We can redistribute the benefits of the market for some, destroying those benefits for others. Every interference in the market represents a trade-off. That trade-off may occasionally be worthwhile. But trusting an elite caste to determine which groups ought to benefit at the expense of other groups cuts directly against a rights-based society—and in the end, against both individuals and the success of the nation.

Markets are natural outgrowths of human nature, and natural rights. You own yourself, and you own your labor—and no one has the right to remove that labor from you for the good of the collective without just compensation. The success of free markets rests precisely on the fact that it reflects truths about human nature: about our desire to create, to acquire, to own, to control the environment around us. As with other utopian schemes, the desire to change human nature by overthrowing individual rights is doomed to both failure and oppression.

By adopting the language of exploitation, "common-good conservatives" embrace leftist rhetoric in pursuit of socially conservative ends. And by suggesting that it is the job of an economic system to build family units, "common good conservatives" commit a serious category error—and in doing so, fall wholesale into Marxist materialism, the notion that all human activity can be explained by economic systems. Economic systems that work create prosperity. Spiritual and cultural systems that work create functional families. For thousands of years, people living in the worst sort of economic privation married and had families and went to church. Then, with the advent of free markets, family structure remained for a century and a half. Then, with the advent of left-driven social change, families collapsed. To attribute the latter to the former is to hitch the future of social fabric to an absolute reduction in human prosperity—creating not only a wrongheaded admixture, but a dangerous one. And to then ladle in an extra helping of demagoguery makes the brew even more toxic.

But Carlson goes even further. He openly embraces the government planning of Elizabeth Warren. After Warren presented a top-down, government-run plan that amounted to, in her words, "aggressive intervention on behalf of American workers," Carlson quoted it liberally, homing in on her phraseology: "Sure, these companies wave the flag, but they have no loyalty or allegiance to America. . . . Politicians love to say they care about American jobs. But for decades, those same politicians have cited 'free market principles' and refused to intervene in markets on behalf of American workers." Carlson then added, "Was there a single word that seemed wrong to you? Probably not . . . She sounds like Donald Trump at his best."[82] Carlson has also attacked investors like Paul Singer as "vulture capitalists" for buying companies and moving or downsizing them in order to make them profitable. At no point has Carlson—or any "common-good conservative"—explained just how they plan on maintaining the prosperity of a free-market economy while ruling out the mechanisms that allow free markets to operate.

Our culture of economic rights creates not only dynamism, but more dynamic Americans. Economic freedom encourages us to forge forth, to create, to follow our dreams, to becomes entrepreneurs, to compete and overcome. Carlson has expressly argued *against* that culture. When I interviewed him in 2019, he maintained that it was unjust for Americans to be expected to move from the towns where their grandparents had lived and died, and suggested outlawing self-driving trucks in order to maintain trucking jobs. Many Americans apparently agree: people now move less than they ever have, despite the ease of travel. In the 1950s, one in five Americans moved every year; today that number is under one in ten.[83] Meanwhile, the American start-up rate is dropping precipitously, particularly among young Americans.[84] And young Americans, more and more, are embracing the notion of a large government that covers social services—and not coincidentally, demonstrating more and more antipathy to economic freedom.

A culture of economic freedom is being replaced by a culture of economic expectation. And a culture of economic expectation inevitably results in a culture of economic tyranny.

CONCLUSION

Step two in destroying America is undermining our culture of rights in favor of a culture of protection by government. Once again, this means supplanting trust in each other individually and societally for a culture of roving virtue-signaling mobs, advocating top-down government controlled by some at the expense of others. And that culture makes every hot issue feel like a breaking point, and every political decision feel like life-or-death.

Our Unionist culture used to be a free-flowing, chaotic, raucous place, motivated by the understanding that individual rights must predominate, and must be balanced by our social institutions. But that balance—individual rights, protected by government, and social duties, promoted by social institutions—has been overthrown by the alternative Disintegrationist culture. That Disintegrationist culture promises instead individual guarantees provided by government—guarantees of positive "freedom" from want and fear, rather than negative freedoms of speech and adventure—and duties created by the mob, dictated by a cultural elite at best indifferent to and at worst openly opposed to the institutions of church and family.

The Disintegrationist culture is winning because it is a culture of expectation rather than a culture of duty. John F. Kennedy once suggested that Americans ask not what their country could do for them, but rather what they could do for their country. That formulation was half-wrong—Americans built a culture in which action on behalf of family and business and community and country was supposed to be, in essence, identical. But the formulation was half-right, too: Americans asked only for the opportunity for freedom.

The Disintegrationist culture dismisses freedom on behalf of security. It enervates rather than energizes; it urges a spirit of solicitation rather than a spirit of enterprise. In the end, the Unionist culture was a culture that relied on the ambition of individuals, the strength of social institutions, and the desire for freedom the founders believed characterized the American experience. The Disintegrationist culture relies on the risk-averse nature of individuals, the weakness of social institutions, and the desire for security over liberty.

The Disintegrationists make a better bet—that human beings are venal, greedy, and self-serving, that human beings would rather invade the freedoms of others than absorb the pain of others' freedom. That bet, however, will end in tragedy. Our culture is like a marriage: it will succeed only when the partners respect one another, and when the partners expect more from themselves than from their counterparts. If Disintegrationist culture wins, the result will not be a new culture, but a national divorce.

CHAPTER 5

THE AMERICAN HISTORY

American history has always been contentious ground.

The roots of American philosophy and culture can be found in our common history. But our history is replete with complexity: with heroism and vision, with idealism and bravery . . . but with victimization and cruelty, too. Because America's history, like all history, contains shades of gray—because no person, no country, no nation can be universally good or universally evil across time and space—it is easy to mischaracterize the general thrust of American history. Viewing American history as a pointillist series of seemingly random evils and goods misses the point—it stands too close to the picture. Draw back, and the picture becomes clear. When we look at that broader picture—when we examine the histories of other nations and other countries, when we observe the suffering and barbarity expressed virtually universally—it is easier to see that American history is far more light than dark. In fact, it is fair to say that while America has participated in evils common to all humanity, America has brought to humanity immense *good* that is entirely unique. Without America, the world would have fallen to tyranny long ago; without America, the beauty of individual rights would have been completely subsumed by the collective long ago; without America, the notion of a multiethnic democracy would have collapsed long ago.

The Unionist history of America is a shared creed-based history of tragedy and triumph. That history is the story of American exceptionalism: why America is different. And that story is the story of our philosophy and culture, rooted in foundational ideals, expressed more and more broadly over time. It is the story of an American moral immune system, challenged by a series of brutal viruses that have sometimes brought the body politic near death, but ultimately emerging victorious, again and again. Each time the American immune system defeats a virus, it is strengthened.

In the view of Disintegrationists, American history is the fruit of a poisonous tree: a history that sprang from a corrupt seed, planted its ugly roots in foreign soil, and then grew, tentacle by cancerous tentacle. In the view of Unionists, by contrast, American history began with a uniquely magnificent seed. Over time, the tree that sprang from that seed grew tall, strong, true, unwarped. Yes, that tree had cancerous branches that required pruning—pruning that often came close to killing the tree itself. But over time, the tree was made healthier by that pruning. And the tree was strengthened by more and more Americans grafting their stems to America's roots, thickening the trunk, making the entire tree more durable and more stable.

None of this means—again—that American history doesn't have a dark side, most obviously in the treatment of Native Americans and black Americans. But it is to argue that in America, goodness and strength overcome horror—not in every instance, not immediately, but over time, thanks to adherence to founding philosophy and culture. The story of the United States is the story of the American founding—a founding steeped in freedom and liberty, in a common creed, expressed via the Declaration of Independence and Constitution of the United States; the story of a nation struggling to overcome its original sin, slavery, and sacrificing hundreds of thousands of men to do so; the story of a country gradually extending the rights guaranteed in our founding documents to black Americans, women, and other minorities. This is the story of the nation that freed the world of Nazis and communism, that powered the economy of the world and raised billions from poverty. This is a story marred by shameful acts—the vicious treatment of Native Americans, the enslavement of blacks, the exclusion of Chinese and Jews—but a story in which horror is overcome by goodness and strength.

The Unionist history argues that America was always great. In short, Unionist history argues in favor of three key principles: first, that America was born with glorious ideals; second, that America has been united by those ideals more and more universally over time, rather than divided by sectarian interest, and that adherence to those ideals has been at the center of America's progress; and third, that the world has benefited from America's power and greatness.

And those who have fought most brilliantly for American unity have agreed. On March 4, 1861, Abraham Lincoln stood on the steps of the still-uncompleted Capitol in Washington, D.C., to deliver his First Inaugural Address. There, in the shadow of a looming war that would cost America more lives than all her other wars combined, Lincoln called upon Americans to remember that they were, after all, brothers. "We are not enemies, but friends," he said. "We must not be enemies. Though passion may have strained it must not break our bonds of affection. The mystic chords of memory, stretching from every battlefield and patriot grave to every living heart and hearthstone all over this broad land, will yet swell the chorus of the Union, when again touched, as surely they will be, by the better angels of our nature."

Memory.

A people must share a history if they are to be a people. And as the American people grew, encompassing new groups, our history changed, too. But that history does belong to all of us, which is why we must remember it. Either we share that past, or we let it divide us.

In order to understand America's current divides over history, we must briefly review the traditional Unionist view of American history. That history isn't black-and-white; it doesn't cover up America's flaws. But it does rest on the foundations of American philosophy and culture, which means that American history, given the broad span of time, is self-correcting—and that when we hold true to our principles, we are true to both our history and our decency.

THE REVOLUTIONARY GENERATION: GLORIOUS BEGINNINGS

The history of America obviously spans centuries further back than the history of the United States. Natives populated America for millennia prior to the West's discovery of America; Christopher Columbus discovered the West Indies in 1492; European countries began divvying up the so-called New World nearly immediately; British attempts to colonize North America began in 1587, with the first successful British settlement taking place in 1607 in Virginia, and Puritan refugees settling at Plymouth Rock in 1620. For purposes of Unionist thinking, however, the true history of the United States begins with the revolutionary generation—with the *United States*.

The revolutionary generation was populated by extraordinary men: George Washington, Thomas Jefferson, John Adams, Benjamin Franklin. These were well-read, brilliant men who were preoccupied with the relationship between morality and government, between liberty and duty. Their belief systems provided the basis for the Declaration of Independence and Constitution of the United States, the former of which was an attempt to state the permanent truths of American philosophy, and the latter of which was an attempt to embody those truths in a workable form of government. The Declaration of Independence restated uniquely American but supposedly self-evident truths:

> [T]hat all men are created equal, that they are endowed by their Creator with certain unalienable Rights, that among these are Life, Liberty and the pursuit of Happiness; That to secure these rights, Governments are instituted among Men, deriving their just powers from the consent of the governed; That whenever any Form of Government becomes destructive of these ends, it is the Right of the People to alter or to abolish it, and to institute new Government, laying its foundation on such principles and organizing its powers in such form, as to them shall seem most likely to effect their Safety and Happiness.

The founding fathers were, of course, products of their time. This meant that from the outset, America was plagued with the contradiction between a universal morality and a time-bound failure to fulfill that same morality. At the time of the American founding, slavery was commonplace across the planet. According to Henry Louis Gates, over the period 1525 to 1866, 12.5 million Africans were shipped to the New World in the transatlantic slave trade. They endured the unthinkably horrible Middle Passage; many died in transport from disease, starvation, or suicide. Of those 12.5 million, approximately 10.7 million arrived in the New World. A grand total of 388,000 landed directly in North America.[1] This is not to excuse a single kidnapping, enslavement, or sale of a human being—each one was a crime against man and God. It is to point out that the United States was certainly not unique at the time of the founding in allowing slaveholding.

Britain was not, either. Britain had not outlawed slaveholding or the slave trade in its colonies; Britain would only outlaw the slave trade in 1807 and emancipate slaves throughout the British empire in 1833.[2] In fact, the original Declaration of Independence included a provision by Thomas Jefferson blaming the British government for the importation of slaves into America:

> He [King George III] has waged cruel war against human nature it-self, violating its most sacred rights of life and liberty in the persons of a distant people who never offended him, captivating and carry-ing them into slavery in another hemisphere or to incur miserable death in their transportation thither. . . . Determined to keep open a market where men should be bought and sold, he has prostituted his negative for suppressing every legislative attempt to prohibit or to restrain this execrable commerce [that is, he has opposed efforts to prohibit the slave trade].

The provision was struck at the behest of southern delegates. But it is perfectly obvious that the Declaration of Independence and its credo that "all men are created equal" was not meant to exclude slaves philosophically; it was meant to encompass everyone. That is why ex-slave and second founding father Frederick Douglass described the "great principles" contained by the Declaration of Independence. Douglass stated, "The signers of the Declaration of Independence were brave men. They were great men too—great enough to give fame to a great age. . . . They believed in order; but not in the or-der of tyranny. With them, nothing was 'settled' that was not right. With them, justice, liberty and humanity were 'final'; not slavery and oppression." Douglass's great cry for freedom arose from his invoca-tion of those founding principles: "Are the great principles of political freedom and of natural justice, embodied in that Declaration of Independence, extended to us?"[3]

Many of the founding fathers opposed slavery morally, even if they were willing to countenance its continuation in order to maintain the Union. This wasn't because they valued the Union more than their values. They believed the Union was the best way to protect and expand the rights they valued the most.

While some forty-one of the fifty-six founding fathers who signed the Declaration of Independence held slaves,[4] many of them spoke out clearly against slavery, including many of the slaveholders themselves. John Adams stated, "every measure of prudence, therefore, ought to be assumed for the eventual total extirpation of slavery from the United States. . . . I have, through my whole life, held the practice of slavery in abhorrence."[5] Benjamin Franklin became the head of the Pennsylvania Abolition Society and stated, "That mankind are all formed by the same Almighty Being, alike objects of his care, and equally designed for the enjoyment of happiness, the Christian religion teaches us to believe, and the political creed of Americans fully coincides with the position. . . . [We] earnestly entreat your serious attention to the subject of slavery—that you will be pleased to countenance the restoration of liberty to those unhappy men who alone in this land of freedom are degraded into perpetual bondage and who . . . are groaning in servile subjection."[6] John Jay, the first chief justice of the Supreme Court, stated, "That men should pray and fight for their own freedom and yet keep others in slavery is certainly acting a very inconsistent, as well as unjust and perhaps impious, part."[7] Jay and Alexander Hamilton were among the men who created the New York Manumission Society in 1787. Gouverneur Morris said slavery was a "nefarious institution" and described "the curse of heaven on the States where it prevailed."[8]

Even slaveholders were aware of the moral blot that slavery represented. George Washington stated, "I can only say that there is not a man living who wishes more sincerely than I do to see a plan adopted for the abolition of it [slavery]."[9] Many of the founding fathers released their slaves, including Washington (he was unable under the laws of Virginia to free the slaves of his wife's children).[10] Jefferson, famously a slaveholder himself and the father of children by a woman he kept in bondage, Sally Hemings—the half sister of his first wife— spoke repeatedly about the immorality of slavery; he called it a "moral blot" and a "hideous depravity."[11] In 1778, Jefferson even introduced a bill to ban the importation of slaves into Virginia, hoping for slavery's "final eradication."[12]

In the northern states, of course, slavery was far less prevalent, and legislators moved early to curtail it. In 1777, Vermont banned adult slavery in its constitution;[13] in 1780, Pennsylvania passed the Gradual Abolition Act, designed to end the practice eventually (although slaves before 1780 were grandfathered in under the bill);[14] in 1783, ruling in accordance with the 1780 state constitution, Massachusetts effectively abolished slavery;[15] Connecticut and Rhode Island and New Hampshire and New York and New Jersey all moved against slavery legislatively by 1804.

The Northwest Ordinance, signed by George Washington in 1787, banned slavery in new territories, which would become the states of Ohio, Indiana, Illinois, Michigan, Wisconsin, and Minnesota. As Lincoln also noted in his 1860 Cooper Union speech, twenty-two of the thirty-nine framers of the Constitution had voted on the Northwest Ordinance; twenty voted in favor of the legislation; another framer, George Washington, signed it.[16] By the time of the Constitution, 60,000 free black Americans lived in the United States; by 1830, there were 300,000.[17]

The Constitution of the United States—a compromise document designed to maintain the Union, despite serious internal disagreements—banned importation of slaves beyond 1808. The infamous three-fifths clause, which counted slaves as three-fifths of a human being for purposes of congressional apportionment, was actually designed to *prevent* increased power in slaveholding states—counting slaves fully would have boosted southern representation in Congress without increasing the number of actual voting citizens. Yes, it was a compromise—and a compromise ardently opposed by slavery opponents like Morris, who railed, "the inhabitant of Georgia and South Carolina who goes to the Coast of Africa, and in defiance of the most sacred laws of humanity tears away his fellow creatures from their dearest connections and damns them to the most cruel bondages, shall have more votes in a Government instituted for protection of the rights of mankind, than the Citizen of Pennsylvania or New Jersey who views with a laudable horror, so nefarious a practice."[18] But if the South had had her way, slaveholders would have had the benefit of fully counting slaves for purposes of apportionment. And had the Union not held, the South would have been its own independent nation—a nation in which slavery would likely have expanded and grown vastly beyond even its evils in the United States, for a far longer period of time.

The Constitution of the United States made no overt reference to slavery, avoiding enshrining slavery in federal law; as Lincoln would later say, "Thus, the thing is hid away, in the Constitution, just as an afflicted man hides away a wen or a cancer, which he dares not cut out at once, lest he bleed to death; with the promise, nevertheless, that the cutting may begin at the end of a given time."[19] James Madison, the father of the Constitution, agreed: he wrote that it would be wrong to place in the Constitution any admission of "the idea that there could be property in men."[20]

Nonetheless, the reality of slavery was not eviscerated by founding ideals. While many of the founding fathers believed that slavery would wither on the vine as time progressed, the creation of the cotton gin made slavery far more economically viable than it had been before, exploding the demand for slaves throughout the American South. In 1781, the total American population was 3.5 million; about 575,000 were slaves; by 1801, those numbers were 5.3 million and 900,000, respectively; by 1830, 2 million slaves were held in bondage in the United States, out of a total American population of 12.8 million.[21] Jefferson was quite correct when he predicted that slavery would eventually end in cataclysm, with slavery judged morally evil by God Himself: "Indeed I tremble for my country when I reflect that God is just: that his justice cannot sleep for ever: that considering numbers, nature and natural means only, a revolution of the wheel of fortune, an exchange of situation is among possible events: that it may become probable by supernatural interference! The almighty has no attribute which can take side with us in such a contest."[22] The failure of Americans to live up to the founding ideals of the Declaration of Independence would bear bloody fruit.

GO WEST, YOUNG MAN!

In the decades after the founding of the United States, the single broadest trend that predominated was the spread of American presence across the continent. This trend was underscored by four main themes: the singular bravery and entrepreneurship of Americans, crossing vast distances to plant roots in distant places and build in support of their families; the increasing ability of the American federal government to gain vast swaths of territory through treaties with major foreign powers; the violation of the rights of Native Americans by the federal government; and the continuing festering of the moral conflict over slavery between North and South.

Today we tend to dismiss the accomplishments of the millions of men and women who crossed mountains and forded rivers seeking new opportunities, without the protection of a vast government designed to shield them from violence and privation. But these were hardy human beings, leaving well-established communities to build new lives. The Americans who traversed the continent, building towns and farms in their wake, were rough and ready. They were also tempted by the seemingly virgin land beyond the Appalachians, which brimmed with beauty and agricultural possibility. By the turn of the nineteenth century, there were approximately 600,000 Native Americans in what would become the continental United States;[23] this was a marked decline from the millions who had populated the continent prior to the epidemics of disease experienced by indigenous peoples upon contact with Europeans. The land area of the United States in 1800 measured 864,746 square miles, with a population density of 6.1 people per square mile;[24] the total land area of the continental United States is approximately 3.5 million square miles, which means that Native American population density was, at that time, 0.17 persons per square mile. The Census Bureau historically considered any area "unsettled" with a population density of fewer than 2 people per square mile.[25] In general, the pioneers were not wrong to cast their eyes westward—although, as we will see, this population movement had dire consequences for Native Americans.

The migration of Americans across the continent had been foreseen by Thomas Jefferson, who saw it as a bulwark against European predations: "we shall form to the American union a barrier against the dangerous extension of the British Province of Canada and add to the Empire of Liberty an extensive and fertile Country thereby converting dangerous Enemies into valuable friends."[26] Jefferson recognized—as did everyone else—that major foreign powers had staked their own claims to huge portions of North America. In 1800, the United States was bordered in the north by British Canada; in the south by Spanish Florida; in the west, by French Louisiana. Spain claimed the vast territory of Mexico, California, and Texas, while Britain and Spain argued over the Pacific Northwest; Russia occupied Alaska.[27]

Early American citizens were deeply concerned about the realistic possibility of broad-based European-style war on the continent. The blood that had soaked the soil of Europe, they believed, could only be stanched by American dominance of the Western Hemisphere. This was no idle thought. European powers weren't willing to give up their claims easily: war in the early days of the republic between proxy groups at the borders was commonplace and vicious. Foreign powers were willing to use aggression of their own in order to maintain their territorial holdings—foreign powers routinely armed Native American tribes, who saw in the United States a threat to their traditional territorial preserves. Jefferson saw the creation of that "Empire of Liberty" as inevitable, and as a preemptive measure against pushback from neighboring entities:

> [W]e have reason to believe that a very extensive combination of British and Indian savages is preparing to invest our western frontier. To prevent the cruel murders and devastations which attend the latter species of war and at the same time to prevent its producing a powerful diversion of our force from the southern quarter in which they mean to make their principal effort and where alone success can be decisive of their ultimate object, it becomes necessary that we aim the first stroke in the western country and throw the enemy under the embarrassments of a defensive war rather than labour under them ourselves.[28]

This philosophy led Jefferson to break with his own interpretation of the powers of the federal government in order to purchase the

Louisiana territory from Napoleon; it led to the federal government looking the other way when General Andrew Jackson pursued aggression in Florida that ended with Spain ceding the territory to the United States; it led to the famous Monroe Doctrine—the American policy designed to bar European interference in the Western Hemisphere, in which President James Monroe informed Congress that "the American continents, by the free and independent condition which they have assumed and maintain, are henceforth not to be considered as subjects for future colonization by any European powers";[29] it led to the sponsorship of the Mexican-American War by President James K. Polk, and the annexation of Texas.

America's expansion did not come without heavy human cost, or without extraordinary cruelty and dishonesty.

While the European powers saw the Native Americans as temporary allies against their great-power adversaries, and while many Americans saw Native Americans as obstacles to expansionism, Native Americans weren't merely proxies for foreign powers or obstacles— they were inhabitants of the land long before any foreign powers had located the New World. And they were fighting for their tribal territory. Many Native Americans chose instead to leave the tribal life; as historian Paul Johnson points out, "Many settled, took European-type names, and, as it were, vanished into the growing mass of ordinary Americans." But Native American tribes would not simply cede land they saw as a tribal inheritance. In fighting the Americans, they employed brutality and cruelty, to be certain, but they correctly foresaw that European dominance on the continent would mean their marginalization at best, and destruction at worst. The conflict between Americans and Native Americans escalated precipitously during the War of 1812, when the Creek tribe decided on war with the Americans, with Tecumseh telling a confederation of tribes, "Let the white race perish! They seize your land. They corrupt your women. They trample on the bones of your dead! Back whence they came, upon a trial of blood, they must be driven! Back—aye, back to the great water whose accursed waves brought them to our shores. Burn their dwellings—destroy their stock—slay their wives and children, that the very breed may perish! War now! War always! War on the living! War on the dead!"[30]

The response to Creek massacres came from General Andrew Jackson, who was just as brutal as Tecumseh. In 1814, Jackson crammed down a treaty on thirty-five chiefs, who were forced to cede 23 million acres to the federal government.[31] As president, Jackson would falsely offer a "just, humane, liberal policy" to Native American tribes—their removal west of the Mississippi River. But in reality, that policy would amount to the forced removal of some 60,000 Native Americans, in violation of their rights, and without proper provisioning. The resulting "trail of tears" left thousands dead.[32] America's treatment of Native Americans would continue to remain brutal for decades.

This behavior was morally inexcusable. It was also in clear opposition to the philosophy of equality before the law that this country was founded on. The expansion of the United States came at the expense of basic human decency. And as with slavery, it would be remiss to fail to place America's territorial expansion in historical context. This is not an excuse; it is a reality. America was *not* exceptional in its expansion. Such expansion was also taking place around the globe at a rapid rate—and war over human migration has been a consistent feature of life. As Johnson notes, during the period of American expansion, "Europeans . . . were moving into the former hunting grounds of the world's primitive peoples in five continents."[33]

It should also be noted that conflict was not foreign to American shores—that intertribal warfare was common between Native American tribes long before the advent of the United States. That warfare was often brutal and terrible; thanks to lack of written history, mass graves tell the story.[34] According to Lawrence Keeley, a professor of archaeology, a survey of western North American Indian tribes and bands found that just 13 percent had not raided or been raided more than once per year. Mutilation of corpses, such as scalping, had been common for centuries (and was later adopted by Westerners as a cottage industry). Keeley observes acidly, "Were there never epidemic diseases before Western contact? . . . Were there never population movements or expansions before civilization?"[35]

The United States' expansion was also perfectly predictable, and well in line with history and global practice of the time. Tocqueville considered America's expansion across the continent "certain," well before that objective had been achieved: "they will spread from the shores of the Atlantic ocean to the shores of the South Sea."[36] And Tocqueville foresaw the continuation of the American assaults on Native American tribal existence: "From whatever side one views the destiny of the natives of North America, one sees only irreparable ills."[37]

With the expansion of the United States' territory, the question of slavery grew larger and larger in the American mind. The founders believed they had put slavery on the road to extinction, but the invention of the cotton gin radically changed the economics of slavery for the South. By the time of the Civil War, some 4 million slaves were held in the United States, held by just under 400,000 slaveholders.[38] As Alan Greenspan points out, "by 1861, almost half the total value of the South's capital assets was in the 'value of negroes.' . . . This rapidly expanding industry rested on foundations of unfathomable cruelty." The cotton industry, powered by human enslavement, allowed slaveholders to live large despite fostering a backward economy that demeaned education and industry—factors that would lead to the South's defeat in the Civil War.[39]

Thanks to both the South's total reliance on slavery and the growing antislavery sentiment of the North, the arguments regarding slavery began to change. Southern advocates began to argue the morality of slavery—particularly John C. Calhoun, a virulent white supremacist—and insisted that the federal government enshrine slavery for all time, force free states to return escaped slaves, and allow for the expansion of slavery into new territories. The positive argument for slavery was actually a new development—before the rise of abolitionism, even slaveholders had tolerated but derided the immorality of slavery. Calhoun infamously stated that the key tenet of the Declaration of Independence, that "all men are created equal," produced "poisonous fruit"—such as the argument that slaves ought to be free.[40]

Calhoun, like most southerners, believed that banning slavery in new territories would eventually allow the government to ban slavery outright, through the means of the legislature. With each new state applying for admission to the United States, therefore, the federal government was forced to decide—and in essence, the North and South were forced to negotiate—the question of slavery's extension. This led, for example, to a hard-fought debate about whether the United States ought to go to war with Mexico on behalf of the Republic of Texas: Abraham Lincoln, for one, opposed the war, as well as annexation of Texas, on the basis that he did not want to see slave power increased. It also led to a bloody battle in Kansas, where two new state governments were declared—one pro-abolitionist, one pro-slavery. The seeds of the Civil War, planted centuries before in the importation of African slaves to the United States, were now about to sprout.

A MORE PERFECT UNION

The issue of slavery came to a head with the election of Abraham Lincoln in 1860. Before Lincoln's election, repeated attempts to come to some sort of compromise that would preserve the United States while limiting the scope of slavery had been attempted: the Missouri Compromise of 1820, which admitted Missouri to the Union as a slave state and Maine as a free state, also banning slavery from the lands of the Louisiana Purchase, also prohibiting slavery north of the Mason-Dixon Line; the Compromise of 1850, which admitted California to the United States as a free state, but mandated one pro-slavery senator, and which set the groundwork for the much-despised Fugitive Slave Law, making northern states party to the recapture of escaped slaves; the Kansas-Nebraska Act, which mandated popular sovereignty to decide on whether a state would be admitted as slave or free, and which promptly sparked open warfare inside the state of Kansas. But with the Supreme Court's decision in *Dred Scott v. Sanford* (1857), in which the Missouri Compromise was ruled unconstitutional and black Americans noncitizens (with Chief Justice Roger Taney adopting the vicious language of Calhoun, declaring slavery to the "benefit" of black Americans), and with the rise of the abolitionist Republican Party, which rejected popular sovereignty, the gap between South and North finally became untenable. Meanwhile, southerners fretted that five free states had been admitted to the United States consecutively—and that this could radically shift the balance of power in the Senate, thus relegating slavery to history. John Brown's attempted slave revolt in Harpers Ferry in 1859 only underscored southern concerns.

And indeed, Lincoln promised to put slavery on the road to extinction. In 1858, Lincoln openly expressed the issue:

A house divided against itself, cannot stand. I believe this government cannot endure, permanently, half slave and half free. I do not expect the Union to be dissolved—I do not expect the house to fall—but I do expect it will cease to be divided. It will become all one thing or all the other. Either the opponents of slavery will arrest the further spread of it, and place it where the public mind shall rest in the belief that it is in the course of ultimate extinction; or its advocates will push it forward, till it shall become lawful in all the States, old as well as new—North as well as South.[41]

Lincoln's victory in the 1860 election precipitated the southern rebellion—a declaration that slavery would be preserved forever in the southern states. Other cross-cutting tensions obviously contributed to the outbreak of the Civil War—economic competition between the free-trading South and the protectionist North, a belief in states' rights against supposed federal encroachment—but slavery lay at the center of the conflict. The new Confederacy's declarations of secession spelled out the rationale for rebellion clearly. South Carolina's declaration blamed "an increasing hostility on the part of the non-slaveholding States to the institution of slavery, [which] has led to a disregard of their obligations, and the laws of the General Government have ceased to effect the objects of the Constitution." For the South Carolinian Confederate leaders, the Constitution had been signed largely to *preserve* slavery, and now the Constitution had been rendered moot by events:

> We affirm that these ends for which this Government was instituted have been defeated, and the Government itself has been made destructive of them by the action of the non-slaveholding States. . . . [T]hey have denounced as sinful the institution of slavery; they have permitted open establishment among them of societies, whose avowed object is to disturb the peace and to eloign the property of the citizens of other States. They have encouraged and assisted thousands of our slaves to leave their homes; and those who remain, have been incited by emissaries, books and pictures to servile insurrection.[42]

The Confederate States of America was not hiding the ball. The new Constitution of the Confederacy mandated that all new states would be slave states.[43]

But because the story of America is not the story of unending tolerance for slavery, the Civil War, the bloodiest war in American history, was fought. More than 600,000 Americans lost their lives. Hundreds of thousands of Union soldiers marched into battle singing "The Battle Hymn of the Republic":

In the beauty of the lilies Christ was born across the sea,
With a glory in His bosom that transfigures you and me;
As He died to make men holy, let us die to make men free!

The original lyrics were even more militant, and name-checked John Brown, the fiery abolitionist militant who led a bloody raid on Harpers Ferry in an attempt to launch a slave uprising: "John Brown's body lies a-moulderin' in the grave, his soul is marching on!"

The Civil War was, in short, a vindication of the founding goodness of the United States. Lincoln did not see the Civil War as a break with founding principles, but as a fulfillment of them. The Declaration of Independence, he would declare over and over, had not been completely performed until slavery ended in the United States. Contrary to the Disintegrationist history, Unionist history vindicates Lincoln's judgment: America, by harking back to her moral roots, progressed.

Lincoln didn't initially stand in favor of emancipation—his mandate was to effectuate union, not to free the slaves. But as the war progressed, it became clear that without solving the issue of slavery, the war would have been fought for nothing. As president, Lincoln issued the Emancipation Proclamation and rammed through the Republican Congress the Thirteenth Amendment. Two months after the amendment passed through Congress, he was assassinated. In 1866, the House of Representatives passed the Fourteenth Amendment. The Fifteenth Amendment, guaranteeing black Americans the right to vote, passed Congress in 1869. Thus slavery ended in the United States, and due process and equal protection of the laws were guaranteed to those who had been held in cruel bondage. Freedom did indeed triumph.

But not for long.

In the waning months of the war, General William T. Sherman issued Special Field Order No. 15, which promised forty acres to freed slaves—in all, 400,000 such acres were to be distributed. Sherman promised the availability of mules provided by the army as well. But Andrew Johnson quickly reversed the order in 1865 and restored the land to former Confederate slave owners.[44]

Nonetheless, after Lincoln's assassination, the radical Republicans took the lead on Reconstruction—and proposed far-reaching plans that would have tremendously accelerated recompense to black Americans. They overrode vetoes from President Andrew Johnson on the Civil Rights Act of 1866, for example, attempting to bar state discrimination, enshrine the ability of black Americans to vote, and bar ex-Confederate officers from holding office. This facilitated the rise of so-called carpetbagging—Republicans traveling South to hold office in the former Confederate states. The election of Ulysses S. Grant forwarded the designs of the radical Republicans—Grant made clear that he would use federal force to enforce federal law. After a failed attempt from anti–radical Republicans to take down Grant in 1872, the radical Republicans finally lost the lead within their own party, to disastrous effect: the Republican presidential candidate Rutherford B. Hayes bargained for the electoral votes of former Confederate states in return for granting withdrawal of federal troops. A new reign of terror against black Americans began, led by the Ku Klux Klan and enshrined in law in Jim Crow regimes. Southern landowners chained black Americans to the land through coercive sharecropping contracts, enforced by state and nonstate violence; blacks who could do so left the South. Four more generations of perversity from a wide variety of political figures would follow—from President Woodrow Wilson, who showed *The Birth of a Nation* at the White House, to FDR, who signed into law the deeply flawed GI Bill that would effectively deny benefits to black Americans, particularly in the South. Redlining was common practice; educational opportunities were barred to black Americans. It would take nearly another century before black Americans were placed in a position of legal equality with their white compatriots in the South.

THE GILDED AGE

The end of slavery catapulted America forward into a new age of economic dominance. The Homestead Act of 1862 handed 160-acre plots of land to those who pledged to cultivate it, incentivizing the spread of Americans across the country—and allowing the purchase of vast tracts of land by those who could build biggest.[45] But it was the rise of modern industry that truly revolutionized America. Greenspan draws the picture well:

> In 1864, the country still bore the traces of the old world of subsistence. Cities contained as many animals as people, not just horses but also cows, pigs, and chickens. . . . By 1914, Americans drank Coca-Cola, drove Fords, rode underground trains, worked in skyscrapers, doffed their hats to "scientific management," shaved with Gillette's disposable razor, lit and heated their houses with electricity, flew in airplanes, or at least read about flights, and gabbed on the phone, courtesy of AT&T.[46]

Wealth increased across every area of American society. Americans became richer than the citizens of any other country—by an enormous margin. By 1914, America's per capita income was $344; in Britain, that number was $244. By 1910, America was responsible for 35.3 percent of all global manufacturing, the highest of any country on earth. America's population also skyrocketed, from 40 million in 1870 to 99 million in 1914; most of that increase came from children of citizens, but one-third of the new increase came from vast swaths of new immigrants.[47] American railways spanned the nation, linking transportation and shipping coast to coast; America built new modes of cross-country communication, making informational transfer easy and cheap.

Much of this was due to the American government's lack of control: America operated on the basis of a gold standard; regulation was lax, and growth was incredibly robust, with the cost of living remaining steady. America's friendly business climate created enormous room for innovation, with geniuses like Thomas Edison and Henry Ford trying and failing and trying again and succeeding. America's natural resources in oil and steel gave the country an inherent advantage— but that advantage could easily have been squandered by a country insisting on top-down economic control. The American government, held in check by constitutional limits, could not effect such control. The Supreme Court opinion in *Lochner v. New York* (1905), which struck down a statute that set a maximum number of hours bakers would be allowed to work, summed up the general tone of the time, and the traditional American view of free enterprise: "It is a question of which of two powers or rights shall prevail—the power of the State to legislate or the right of the individual to liberty of person and freedom of contract."[48] While the Court should not have relied on the due process clause to make its case—that legal perversity would rear its ugly head in future cases, including *Roe v. Wade* (1973)—the economic attitude was reflecting of founding-era viewpoints on the role of the government.

Again, the belief system of the founders provided the framework for explosive growth. And that explosive growth would power the greatest economy in the history of mankind. In time, the entire global economy would come to rely on the engine built by Americans, using the blueprint set by the founding fathers.

Now, the American government did provide massive subsidies and privileges to certain railroad magnates seeking territory to build their lines (although, notably, James J. Hill built the Great Northern Railroad without any help from the government).[49] But the story of American growth during the Gilded Age was the story *not* of corruption, but the story of entrepreneurship. After all, corruption is the common flaw of humanity—but entrepreneurship can only thrive in a climate of freedom. Rockefeller and Carnegie grew up poor. These capitalists took risks, and reaped the rewards. New methods of investment, including the limited liability corporation, were created, and allowed people to invest their money without the risk of their personal *non*-invested assets falling into risk. The stock market was an outgrowth of the new diversification of investment sources. As historians Larry Schweikart and Michael Allen write, "The excesses of the Gilded Age, both outrageous and mesmerizing, concealed, as most people knew, a widespread prosperity generated by the most amazing engine of growth ever seen." Even communist Leon Trotsky had to acknowledge that American living standards were astonishing: in 1917, he wrote, "We rented an Apartment [in New York] in a workers' district, and furnished it on the installment plan. That apartment, at $18 a month, was equipped with all sorts of conveniences that we Europeans were quite unused to: electric lights, gas cooking-range, bath, telephone, automatic service elevator, and even a chute for the garbage."[50]

America's growth did not come without costs. The crowding of Americans into cities created serious health risks as well as environmental degradation. Workers complained that capitalist bosses lived high on the hog while they labored in unsafe work conditions; the nascent American union movement began to grow more and more powerful. Strikes became commonplace. The consolidation of major industries created concerns about monopoly power, and perhaps the stagnation of competitive opportunity for those on the bottom of the economic scale. Muckraking journalists like Ida Tarbell termed the great business magnates of her day "robber barons"—a description that would stick, fairly for some, unfairly for others.

The response to these concerns would break upon the horizon with the Progressive era—the complete rewriting of the bargain between Americans and their government.

THE NEW COLOSSUS

By the turn of the twentieth century, America was a global force. She had expanded her territorial holdings (Secretary of State William Seward negotiated the purchase of Alaska from the Russian government in 1867, and the McKinley administration backed an American business-led coup in 1898 that ended with the annexation of Hawaii) and backed national independence movements that would benefit America at the expense of other great powers (the Spanish-American War ended with Cuba's independence from Spain, finally granted after American intervention in 1902, and the Philippines became an American protectorate in 1898 after the United States sought to prevent other great powers from closing the islands to American navy power—the Philippines would gain independence in 1946). Her economy was far stronger than that of any other nation on the planet.

But the stirrings of discontent could not be quelled for long—and progressive politicians, taking their lead from German progressives who advocated for broad government involvement in the economy, reacted. This was an era of great-thinking leaders—even leaders who had benefited from the freewheeling nature of the Gilded Age began to think that top-down control could work on behalf of the broader American economy. President Theodore Roosevelt led the charge, explaining, "The effort at prohibiting combination (corporate growth) has substantially failed. The way out lies, not in attempting to prevent such combinations, but in completely controlling them in the interest of the public welfare." While calling for a "square deal," Roosevelt *actually* wanted to tear down large and successful businesses in the name of public welfare. In 1910, speaking in Osawatomie, Kansas, TR laid out his fundamental precepts (precepts from which Barack Obama would draw inspiration a century later):

> The absence of effective State, and, especially, national, restraint upon unfair money-getting has tended to create a small class of enormously wealthy and economically powerful men, whose chief object is to hold and increase their power. The prime need to is to change the conditions which enable these men to accumulate power which it is not for the general welfare that they should hold or exercise. We grudge no man a fortune which represents his own power and sagacity, when exercised with entire regard to the welfare of his fellows.[51]

In other words, business would do what government wanted, or government would force it to do so. And TR was willing to overrun all constitutional boundaries—and the boundaries of decency—in order to accomplish his reordering. During a 1902 coal strike, TR threatened to send in the army to resolve the situation, reportedly shouting, "To hell with the Constitution when the people want coal!"[52] TR knew full well that the fulminations of Upton Sinclair about the Chicago meatpacking industry were overwrought nonsense (Sinclair himself was a socialist, and his comrades greeted his book as a piece of successful propaganda), but he didn't care—he admitted that Sinclair was "hysterical, unbalanced, and untruthful"—but still used Sinclair's lies to push through the Pure Food and Drug Act of 1906.[53] Under TR's administration, Congress rammed through the Sixteenth Amendment, legalizing the federal income tax; and the Seventeenth Amendment, removing senators from the control of state legislators and making them directly elected. And Roosevelt aggressively utilized the Sherman Antitrust Act of 1890—a vague document calling for trust-busting—against a variety of businesses, supposedly for "restraint of trade," although the results of that trust-busting did not generally lead to lowered prices in the affected industries.[54]

TR's progressivism paved the way for that of Woodrow Wilson. His philosophy of government was clear: it ought to have no limits. "Government," he explained, "does now whatever experience permits or the times demand." His view of leadership was similarly frightening: "Men are as clay in the hands of the consummate leader." Wilson was no outlier: elements of his philosophy could be found in prominent thinkers from John Dewey to Walter Rauschenbusch, a leader of the so-called Social Gospel movement.[55] Wilson was a full-blooded progressive who believed that government ought to control the economy using a network of executive branch bureaucrats, all in the name of breaking the wealthy. "Property as compared with humanity, as compared with the vital red blood in the American people, must take second place, not first place," Wilson demagogued.[56] Wilson opened a Labor Department, signing into law an act that would create a branch of the federal government specifically designed "to foster, promote and develop the welfare of working people, to improve their working conditions, and to enhance their opportunities for profitable employment." How any of this fell under the rubric of the Constitution was anyone's guess—but it was an ambitious age. In practice, the Department of Labor became—and remains—a compulsory tool geared to the agenda of labor unions. Wilson's pro-union agenda was so strong that he utilized World War I in order to remake the American workplace, recognizing union power to bargain collectively, as well as an eight-hour workday.[57] Wilson also joined the general progressive enthusiasm for eugenics.

America's involvement in World War I—perhaps an inevitable consequence of Germany's attacks on American shipping, perhaps not—helped liberate the continent of Europe from German aggression: the Russian government had already been removed and supplanted with a communist one, and had engaged in a separate peace with Germany. Germany would have been free to turn her full might to the war on the Western Front, but American doughboys helped stanch the bleeding and reverse the tide. Historian Geoffrey Wawro points out bluntly, "Germany would have won World War I had the US Army not intervened in France in 1918."[58]

The war also prompted yet more government interventionism at home. Wilson pursued dissenters with the force of law using the Espionage Act (1917) and the Sedition Act (1918), and propagandized about anyone who opposed any aspect of the war. As Jonah Goldberg states, "Even after the war, Wilson refused to release the last of America's political prisoners, leaving it to subsequent Republican administrations to free the anti-war Socialist Eugene V. Debs and others."[59]

Wilson's "big thinking" left its mark on world affairs, too. He had campaigned in 1916 on the promise to stay out of World War I; that promise certainly gave the German government reason to think that it could win a victory in Europe before any American intervention would be forthcoming. In declaring war, Wilson spelled out his vision of an international society, a world "made safe for democracy." Wilson's vision of morality-based internationalism ran in stark contrast to the realism that had dominated American foreign policy heretofore—Wilson even praised the Russian Revolution, which would end in the decades-long domination of the evil Soviet government.[60]

When the war ended, as men flooded back into the economy, strikes paralyzed major industries across the United States—ushering in the rejection of Wilsonian and Rooseveltian progressivism . . . at least for the moment.[61] The roaring 1920s were a direct response to the era of progressivism that immediately preceded it. Americans embraced free living, a free economy, and a freer culture. Hollywood became a cultural font; the American automobile became the symbol of American prosperity; the phenomenon of air travel became possible; the radio became a method of mass communication. Babe Ruth, Charlie Chaplin, and Walt Disney would be introduced to the world. President Calvin Coolidge explained, "The man who builds a factory builds a temple," and stated, "if the federal government should go out of existence, the common run of people would not detect the difference in the affairs of their daily life for a considerable length of time."

And Coolidge understood that the principles of free economics were rooted in founding principles: speaking on the 150th anniversary of the Declaration of Independence, Coolidge stated:

> If all men are created equal, that is final. If they are endowed with inalienable rights, that is final. If governments derive their just powers from the consent of the governed, that is final. No advance, no progress can be made beyond these propositions. If anyone wishes to deny their truth or their soundness, the only direction in which he can proceed historically is not forward, but backward toward the time when there was no equality, no rights of the individual, no

rule of the people.... We live in an age of science and of abounding accumulation of material things. These did not create our Declaration. Our Declaration created them.[62]

As Greenspan points out, between 1921 and 1929, American gross domestic product (GDP) grew 5 percent per year. Prosperity was widespread, and Americans were eager to get in on it.[63]

All of that collapsed not with the stock market crash of 1929, but with the government response to that crash. By the time the market crashed, President Herbert Hoover was in charge. His response to the crisis was quite progressive: he demanded heavy government interventionism. He held a meeting of industrialists at the White House in which he demanded that they maintain higher wages than could be supported by the market—which increased unemployment. He stuck by the tariffs put in place during the 1920s, which led to a crisis of demand. Unemployment jumped from 5 million in 1930 to 11 million in 1931. In June 1932, Hoover raised the top income tax rate from 24 percent to 48 percent, explaining, "We cannot maintain public confidence nor stability of the federal government without undertaking some temporary tax increases."[64] This interventionism, combined with the tariffs that had been put in place during the 1920s, led to a worldwide crisis that would end with mass unemployment and enormous government across the West—and in Europe, would precipitate the greatest conflict in the history of mankind.

Yet that interventionism became the model for Franklin Delano Roosevelt's New Deal. FDR adopted Hoover's policies, but doubled down on them—FDR brain trust member Raymond Moley admitted, "When we all burst into Washington . . . we found every essential idea [of the New Deal] enacted in the 100-day Congress in the Hoover administration itself."[65] Rex Tugwell, an FDR advisor who expressed his awe for the "power of the collective will" in the Soviet Union,[66] similarly stated, "we were too hard on a man who really invented most of the devices we used."[67]

FDR's overweening state was tyrannical in economic matters: the FDR administration pressured businesses to fly the so-called Blue Eagle, which demonstrated compliance with the National Industrial Recovery Act; FDR's administration compelled businesses to restrict production to drive up prices, and to cave to unions via the new and powerful National Labor Relations Board. He engaged in vast schemes of wealth redistribution and currency manipulation, and used the most demagogic language in order to ram them through. He labeled the 1920s years of "false prosperity" and focused his ire on "speculators," averring, "I believe in individualism in all of these things—up to the point where the individualist starts to operate at the expense of society." But this, of course, meant that individualism had no actual ground to stand upon, since nearly every individual action can be read in opposition to the vaguely defined social interest.[68] FDR even termed wealthy Americans "economic royalty," and suggested that "the privileged princes of these new economic dynasties" had to be confronted by the "Minute Men" of the new era, that "the political equality we once had won was meaningless in the face of economic inequality." Better an activist government, acting in the "spirit of charity," violating the boundaries of the Constitution, FDR stated, than "the consistent omissions of a Government frozen in the ice of its own indifference."[69]

FDR's policies, despite his rhetoric, were a vast failure—at least in terms of helping the American economy. They massively increased the size and scope of government, but lengthened the Great Depression by seven years, according to University of California, Los Angeles economists Harold Cole and Lee Ohanian.[70] As the two professors point out, "there was even less work on average during the New Deal than before FDR took office. . . . Total hours worked per adult in 1939 remained about 21 percent below their 1929 level." Consumption also dropped dramatically as the government implemented price-and-wage controls. FDR's administration deliberately violated the laws of competition by cartelizing major industries, allowing major businesses to collude in raising prices in order to raise wages for workers. As with all economic policies designed to benefit a few, the cost was borne by the many.[71] Betrayal of founding principles carried with it economic disaster.

But FDR's policies *did* accomplish one great success: they convinced Democrats that class warfare was effective politics, that government was the solution to every problem, and that a vast restructuring in the very framework of American rights was necessary.

SAVIOR OF THE WEST

Whatever the sins of FDR's economic policies, however, they paled in comparison to the Soviet centralization of economics, which ended in the deaths of tens of millions; whatever the sins of FDR's centralization of power, they paled in comparison to the Nazi centralization of force, which ended in the deaths of millions. And indeed, when the Nazi regime threw off the shackles of the Versailles Treaty, both the United States and Europe dithered. By the time the West awoke to the dangers of a remilitarized Germany and an aggressive Soviet Union, World War II had been launched, and continental Europe was under the control of tyrants from Italy to Finland.

America, lying across the seas, remained the great hope of the West. America had taken an isolationist line during the Great Depression, focusing on its internal economic disaster; America had slashed its own military to the point where in 1939, America's military was ranked nineteenth in size, just behind Portugal and just ahead of Bulgaria.[72] Yet America was widely and correctly perceived as a sleeping giant. Upon taking office as prime minister of Great Britain upon the fall of France, Winston Churchill pledged to fight upon the beaches and on the shores against Nazi tyranny—and promised his people that the empire would fight on until, "in God's good time, the New World, with all its power and might, steps forth to the rescue and liberation of the old."[73] If ever there was a time for Jefferson's "empire of liberty" to stand forth and defend its eternal principles, now was the time.

Indeed, the moral conflict became clearer each day in the United States. In December 1940, a year before Pearl Harbor, FDR warned, "Never before since Jamestown and Plymouth Rock has our American civilization been in such danger as now. . . . The experience of the past two years has proven beyond doubt that no nation can appease the Nazis. No man can tame a tiger into a kitten by stroking it. There can be no appeasement with ruthlessness. There can be no reasoning with an incendiary bomb. We know now that a nation can have peace with the Nazis only at the price of total surrender." In response, he called for a "great arsenal of democracy."[74] That arsenal would be brought to bear after the Japanese sneak attack on Pearl Harbor. Churchill noted in his memoirs that he knew, at the moment that Japan had attacked America, that "we had won after all!"[75]

And America would indeed emerge victorious. Again, as in all wars, this would not come without domestic cost—most obviously and egregiously, the government decision to intern 120,000 people of Japanese ancestry, most of them American (in December 1944, FDR ordered the suspension of his executive order). But without America's intervention in World War II, the likelihood of either Nazi victory or Soviet domination of the entire European continent in the case of Nazi defeat would have been near certain. The Holocaust would certainly have continued for longer, and resulted in the deaths of remaining European Jewry. The Japanese dominance of the Pacific would have continued, and its tyranny metastasized. America's development of a nuclear weapon for use at Hiroshima and Nagasaki, far from contributing to a generalized slaughter, was used specifically and sparingly, ending the war more quickly, and setting the stage for an age of nuclear deterrence that would radically reduce the number of war casualties over the course of the next decades.

America's great goodness didn't stop with the end of World War II. In the aftermath of the war, America represented the sole bulwark against both Soviet domination and the return to a dark economic age. America had been untouched on the home front—the only major power not to experience significant destruction within its own borders. America produced 42 percent of *all* manufactured goods, 80 percent of its automobiles, 57 percent of its steel.[76] Western Europe lay in ruins; the Soviet Empire spanned from Siberia to East Germany. The United States remained the West's wall against Soviet domination, against the "iron curtain" described by Churchill. By 1948, the United States had hatched a plan to provide more than $15 billion to rebuild the ruined continent of Europe, up to 5 percent of American GDP at the time; George Marshall, the father of the so-called Marshall Plan, explained, "Our policy is not directed against any country, but against hunger, poverty, desperation and chaos."[77] The United States transformed Germany and Japan into democratic republics, and helped fund the rebuilding of France, Britain, and Italy. The United States also took the leading role in the formation of the North Atlantic Treaty Organization (NATO), an alliance of democracies seeking mutual defense against Soviet aggression. When the Soviets threatened to close off Berlin in 1948, shutting down highways, railroads, and canals in the process, the United States mobilized a yearlong airlift that shipped more than 2.3 million tons of cargo into West Berlin, keeping the city free in the midst of communist East Germany. "We shall stay, period," Truman said. And the United States did. By 1949, the Soviets had ended their blockade.[78]

Meanwhile, the Soviets were actively expanding their global reach to Cuba and South America. In Asia, with Soviet help, communist tyrannies were on the move everywhere from China to Korea. Again, America stepped into the breach. In Korea, the United States sacrificed tens of thousands of brave Americans to guarantee the liberty of those thousands of miles away from America's shores; as President Harry Truman expressed in his 1950 public statement announcing America's commitment to a United Nations action in Korea, "The attack upon Korea makes it plain beyond all doubt that Communism has passed beyond the use of subversion to conquer independent nations and will now use armed invasion and war."[79] The United States would lose some 45,000 men in Korea, but keep South Korea free.

At home, the United States burst loose economically. With men returning from war to a global economy reliant nearly entirely on American productive capacity, the economy boomed. Couples married young and enthusiastically turned out babies, leading to the so-called baby boom. The government pushed home ownership and college education, and the focus of Americans also shifted gradually from manufacturing to services. Huge companies spanned the globe, and dominated industry. But there were systemic problems, too. America could dominate a world ruined by war. But could it dominate foreign competitors after taking into account the bloat associated with fat union contracts, complacent management, government subsidies, and trade protectionism? More important, as Americans began to clamor for a utopian Europeanism, could the economy continue to grow?

THE AGE OF CHAOS

The 1960s began auspiciously. John F. Kennedy, a leader of the new, optimistic generation, was elected president. America's economy was booming; while the Soviets had beat the United States to orbit, Kennedy pledged that America would beat the Soviets to the moon. Kennedy, an ardent anticommunist, also placed American troops in Vietnam, determined to beat off Soviet aggression that could lead to a domino effect in the Far East. And JFK planned ambitious legislation to finally fight back against the powerful brutality of Jim Crow, which had damned black Americans to second-class status for a century after the Civil War.

With Kennedy's assassination, hope turned to rage.

Sometimes, as in the case of the civil rights movement, that rage was not merely righteous, but wholly necessary. Despite the privations of Jim Crow, black Americans had continued to press for their share of the American dream, heroically fighting for the freedoms promised in the Declaration of Independence and then again by Lincoln and the radical Republicans. They fought for their own economic independence, their right to political involvement, their right to liberty. Despite Jim Crow, the black poverty rate fell from 87 percent in 1940 to 47 percent by 1960; as Stephen and Abigail Thernstrom point out, "The proportion of black families who had to survive on incomes below the poverty line was thus cut in half." This was "well before the Civil Rights Act of 1964 outlawed Jim Crow practices in employment, education, and public accommodations. It was also well before the advent of the War on Poverty and various other Great Society programs designed to uplift the poor."[80]

With the civil rights era came the end of legal discrimination in the United States—although that progress had been burgeoning for a long time, and social progress had indeed been taking place even before the effectuation of full legal protections. The Civil Rights Act of 1964 outlawed discrimination both in the public sector and in "public accommodations"; the Voting Rights Act of 1965 guaranteed federal enforcement of the Fifteenth Amendment. Freedom had once again triumphed over slavery and segregation. That didn't mean that racism ended; it didn't. Informal segregation continued—although racism, by every metric, mitigated radically over the coming decades. But the centralizing principle of American history and American life gained unprecedented strength: freedom continued to overcome tyranny. Once again, Unionist history was proved correct: by remembering founding principles and applying them more broadly over time, America had moved forward toward curing her own ills.

But America would simultaneously stray from her founding philosophy and culture—to dire effect. Unfortunately, the chaos of the 1960s was not limited to righteous causes—or to useful action. The second-wave feminist movement moved beyond the rights to vote and work—traditional elements of female equality with broad support from Americans—and toward abortion and "sexual liberation," and fights against the traditional institution of marriage and the biological imperative of childbearing. The second-wave feminist movement argued instead that true liberation could come only in tearing down these crucial, "patriarchal" institutions. Elements of the civil rights movement argued in favor of massive government-provided benefits, even justifying violent direct action, including riots. The Vietnam War was a flashpoint for college students who saw America as a malign influence in the world, despite the fact that America had no imperialist interest in Vietnam, that the Soviets and Chinese were interested in the tyrannical overthrow of an anticommunist government in South Vietnam, and that the fall of Vietnam would indeed result in a domino effect of communist horror throughout the region, most obviously in Cambodia.

The orientation of young Americans toward their own country reversed. Where young Americans had seen the country as a force for good in the world, many young Americans now began to see America as a place of oppression and tyranny. The role of religion in American

public life came under heavy scrutiny, and church attendance plum-
meted.[81] The mechanisms of public information were tilted heavily in
favor of the political left, from television to newspapers.

At the same time, the traditional role of government was overthrown in the grandest manner in the history of the republic. Lyndon
Baines Johnson, inheriting the presidency from the slain Kennedy,
committed himself to entirely rewriting the bargain of Americanism.
From now on, LBJ suggested, government's role would be to guarantee not merely equal opportunity, but equal result.[82] To that end, LBJ
declared the advent of the "Great Society," which would fight a "war
on poverty." That war included an unprecedented increase in the size
and scope of government, from Medicare and Medicaid to extension
of Social Security and disability benefits, from federal housing and
Head Start programs to increasing the size of the Aid to Families with
Dependent Children program. And LBJ also utilized the auspices of
the federal government to hire people to act *against* the government
itself, utilizing protests and strikes as tools. As Amity Shlaes notes,
"by 1971, for the first time, federal spending on what we now call
entitlements—benefits for the aged, the poor, and the unemployed—
outpaced spending on defense."[83] President Richard Nixon only doubled down on his predecessor's programs, expanding the welfare state
even more than LBJ had.

The result of the new, Leviathanic welfare state was tragic: by
severing the founding view of human nature and natural rights from
government, the incentive structure toward individual responsibility
and ambition was completely skewed. The progress of black families
stalled out; by 1970, "progress in reducing black poverty came to a vir-
tual halt," the Thernstroms reported. "The black family poverty rate
was 30 percent in 1970, 29 percent in 1980, and 26 percent in 1995."[84]
But the problem for American blacks was particularly severe, because
the racial gap between black and white ceased to close. As early as

1965, future senator Daniel Patrick Moynihan, then the assistant secretary of labor under LBJ, pegged the problem: blacks were faring "worse, not better," thanks to the breakdown of family structure—a breakdown largely due to the legacy of both slavery and Jim Crow, but exacerbated by government policy.[85]

Overall, the American economy was stalling, too. Poverty rates across racial lines stopped dropping around 1970.[86] In 1971, Richard Nixon fixed prices, wages, salaries, and rents; instituted new tariffs; and removed America from the Bretton Woods Agreement of 1944, thus unmooring the American dollar from the value of gold.[87] American business was no longer competitive on the world stage—by the mid-1960s, American labor costs in the automotive industry, for example, tripled those in Europe, and were quadruple or quintuple those in Japan.[88] Former manufacturing hubs began to empty out. From 1960 to 1975, the Dow Jones Industrial Average was completely stagnant. Only in the early 1980s, as we will see, did the stock market begin its long, unparalleled climb.

The general stagnation that ended the postwar boom was not merely economic. It was societal. Race riots broke out across the nation beginning in the early 1960s: earlier race riots, largely by racist whites against black Americans, had been common just after World War I; the new brand of riots largely occurred among black Americans, ranging from Watts (1965) to Detroit (1968). Vietnam War protests turned violent; the 1968 Chicago Democratic convention turned riotous. Acts of terrorism began to plague American soil: student radicals, including offshoots of some of the student groups actively promoted by LBJ allies in the early 1960s, bombed targets across the country, including the Pentagon. In one eighteen-month period stretching from 1971 to 1972, according to the FBI, the country was hit with approximately five bombings *per day*.[89] Crime rates skyrocketed: according to criminologist Barry Latzer, "Starting in the late 1960s, the United States suffered the biggest sustained rise in violent crime in its history. In some locales, people dreaded leaving their homes at any time, day or night."[90] In 1974, President Richard Nixon resigned; in 1975, America withdrew its final troops from Saigon, and South Vietnam fell to the communists; in 1979, the Soviets invaded Afghanistan and the new Iranian radical Islamic regime stormed the American embassy, taking fifty-two hostages. America's governmental and social institutions were in trouble; America's economy was in trouble; America's foreign policy was in trouble. The election of Jimmy Carter didn't change any of that—it merely underscored the problem, as America fell into the trap of stagflation and the new president lectured Americans about their "crisis of confidence."[91]

THE REAGAN REVOLUTION

The election of Ronald Reagan came as a shock to many in the intelligentsia. But his prescription for America's ills was rooted in years of preparation. The former head of the Screen Actors Guild, former General Electric spokesperson, and former governor of California had been well versed in the threats of communism; he recognized the unique evil of the Soviet Union, which he famously termed an "evil empire," angering hordes of more dovish intellectuals. He also understood clearly that the role of government had been removed from the principles of the founding fathers. As Reagan explained in his First Inaugural Address:

In this present crisis, government is not the solution to our problem; government is the problem. From time to time, we have been tempted to believe that society has become too complex to be managed by self-rule, that government by an elite group is superior to government for, by, and of the people. But if no one among us is capable of governing himself, then who among us has the capacity to govern someone else? . . . It is my intention to curb the size and influence of the Federal establishment and to demand recognition of the distinction between the powers granted to the Federal Government and those reserved to the States or to the people.[92]

This amounted to nothing less than an utter reversal of American movement toward progressivism—a movement that had continued virtually unabated for eight decades, with the brief exception of the 1920s. Reagan's economic agenda was threefold: first, lowering taxes in order to stop punishing businesses for success, and to incentivize investment and growth; second, to relieve regulatory burdens that had hamstrung entrepreneurs; and finally, to stop inflation in its tracks, granting the predictable basis for economic growth missing since the advent of government growth under LBJ. By the end of Reagan's administration, the American economy was booming—America had produced 14 million new net jobs. The Dow Jones traded at just over 900 the month Reagan took office; when he left, it was over 2,200. The financial industry boomed as investment capitalism replaced managerial capitalism. Reagan's greatest failure lay in his inability to rein in federal spending—which has continued to skyrocket to this day.

But Reagan also restored America's place in the world. In 1980, just before Reagan took office, Carter secretary of defense Harold Brown acknowledged that the United States faced a turning point that could be deferred "no longer." Brown averred, "We must decide now whether we intend to remain the strongest nation in the world."[93] That assessment came in the wake of congressional cuts to the defense budget—cuts radically reversed by the Reagan administration and Congress, which increased spending by nearly two-thirds.[94] Reagan committed to building a missile defense system—derided as "Star Wars"—in order to counter Soviet ballistic missile development. Those defense buildups led, in turn, to the collapse of the Soviet Union—a development foreseen by nearly no one just a decade earlier, when the Soviets indeed were often considered to have the momentum in terms of global initiative. When Reagan told Soviet premier Mikhail Gorbachev to "tear down this wall"—the Berlin Wall—in June 1987, he simply wasn't taken seriously by the foreign policy elite. Two years later, the Berlin Wall fell.

This wasn't just a matter of military policy. It was a matter of restoring moral clarity. In the aftermath of the Vietnam War—a war that left too many Americans believing that American foreign policy muscularity was damaging and exploitative rather than liberty oriented—Reagan painted a stark contrast between the Soviet "Evil Empire" and the freedom-loving West. That civilizational contrast was true, and a reminder to Americans that we were, in fact, a shining city on a hill, a beacon of freedom for those suffering all over the globe.

A COUNTRY IN SEARCH OF AN IDENTITY

The United States in the post–Cold War era has suffered from a crisis of identity. America's founding principles had been placed in stark opposition to the Soviet Union for nearly a century; with the collapse of that opposition, America had to decide what she stood for. The result was confusion.

In terms of economics and lifestyle, America continued to boom, relying on principles of free-market economics. The early 1990s drop in crime was the largest and most shocking in American history. Meanwhile, the American economy continued to grow throughout the 1990s, as President George H. W. Bush gave way to President Bill

Clinton, a Democrat who pledged that the "era of big government is over."[95] Indeed, Clinton, despite his vast differences with Speaker of the House Newt Gingrich—author of the "Contract with America," which pushed everything from welfare reform to anti-crime measures to a cut in the capital gains tax—worked to balance the budget and end deficit spending. Free trade, which had already begun to work its magic during the Reagan administration, gained more ground with the creation of the World Trade Organization. Eastern European countries, freed from the heavy hand of communism, began to engage with the market. Even China, a communist country, began to experiment with capitalistic measures. Wealth grew; home ownership grew; the stock market skyrocketed. Clinton's own personal foibles could not dent his popularity due to the strong economy.

On the foreign front, however, America struggled with its new role as unchallenged global hegemon. Was America a global policeman, as in Yugoslavia, or a country focused mainly on self-interested foreign policy, as in Somalia? Clinton's solution was to slash the military budget, leaving America ill prepared for the coming of war on September 11, 2001. The newly elected George W. Bush was forced to cope with a multifront war on terrorism. This led Secretary of Defense Donald Rumsfeld to famously quip, "You go to war with the army you have, not the army you might want or wish to have at a later time."[96] But the question of what America should be was still an open one—and that conflict broke out into the open over the war in Iraq. Was the war about America's war against terrorism? Or was it about human rights? Was it about guaranteeing the sanctity of the United Nations? Or was it about shaping the world toward democracy in Wilsonian fashion? Was it about preventing the Iraqi dictator Saddam Hussein from obtaining weapons of mass destruction? Or was it about freeing the Iraqi people from the evils of dictatorship?

When the Iraq War—which resulted in a quick American victory

over the Iraqi military—quickly turned into a bloody insurgency suppression effort, these questions could no longer be stifled. And when
the American economy—a big-spending, entitlement-based "com-
passionate conservative" economy with the additional element of tax
cuts—collapsed in 2007–2008, America came to a crossroads.

Standing at that crossroads was Barack Obama, a unique political
figure with a vague agenda but a clear appeal. Not only was Obama a
gifted orator, but he signified in his very person—a black American,
child to a white mother and a black father—a rebuttal to the notion
that America was bound to its historic racism for all time. Obama's
ascension to the White House could have been yet another vindication
of the application of America's founding principles—if Obama had
believed in those principles. But, as it turned out, Obama did have a
vision for American government—a vision not dissimilar from that of
Johnson, FDR, or Wilson, at least in terms of government interventionism. That philosophy resulted in unprecedented levels of spending, a massive new entitlement program and a major step toward the
nationalization of the entire health care system, and an attempt to
reorder government ever in favor of federal dominance. On foreign
policy, meanwhile, Obama embraced a bizarre admixture of humanitarian interventionism (Libya) and appeasement-oriented isolationism
(Iran), all amounting to radical slashes in the military. Most tragically,
the promise of racial conciliation that seemed nearly fulfilled at his
election quickly gave way to a new era of racial animus: the Obama
administration became a proponent for identity politics centering on
a narrative of America as an oppressive country, with racism baked
into its DNA. The election of Donald Trump, a massively polarizing
figure in his own right, represented a backlash to Obama's agenda—a
backlash that was almost wholly reactionary, given that it stood less
for the restoration of founding principles than for rejection of Obama's
vision for the country.

CONCLUSION

The story of America is neither wholly good nor wholly bad—but it is far more heroic than tragic, far more representative of justice than injustice. A majority white country that once enslaved black Americans and repressed women regularly elects black Americans and women to high office (in fact, women represent a majority of the electorate, and black Americans routinely outvote their percentage of the population), demonstrating the durability and growth of the uniquely American belief that all men are created equal; a country that fought for its own liberty from British tyranny has liberated billions across the globe in the name of that same liberty, and taken in tens of millions of Americans seeking their own American dream; a country founded in relative wilderness has become the engine of growth that powers the entire world.

The story of America is one of the great stories in human history. America was founded on great principles; America has struggled to live up to those principles, but with each step toward those principles, America has magnified its own greatness. The world is better off for America. We ought to understand the shadows and curses of our history; we ought to understand how history affects the present. But we all ought to understand, most of all, that we are part of the same history, not rivals in a country divided by identity or class. Yet, as we will soon see, Disintegrationists have reworked American history to do just that.

CHAPTER 6

DISINTEGRATING AMERICAN HISTORY

Just a few days before Abraham Lincoln gave his First Inaugural Address on March 4, 1861, another American with a long list of patriotic credentials gave his own inaugural address. On February 18, 1861, Jefferson Davis stood in the magnificently columned capitol building in Montgomery, Alabama, and accepted his proclamation as the first president of the Confederate States of America. Davis was born in Kentucky and named after Thomas Jefferson; Davis graduated from West Point and served as a personal guard for the American Indian Sauk leader Black Hawk after Black Hawk was captured by American forces. He married the daughter of his old commander, Zachary Taylor, who would become president of the United States, and served in the Mexican-American War, being wounded at the Battle of Buena Vista. He was eventually appointed senator from Mississippi, served as secretary of war under Franklin Pierce, and then returned to the Senate.

Davis was actually an ardent antisecessionist in the years immediately before the Civil War. In 1858, Davis spoke at Faneuil Hall in Boston, where he advocated *against* secession. "[I]f you have any hope connected with the liberty of mankind, if you have any national pride in making your country the greatest of the earth, if you have any sacred regard for the obligation which the acts of your fathers entailed upon you—by each and all of these motives you are prompted to united and earnest effort to promote the success of that great experiment which your fathers left it to you to conclude," Davis passionately intoned.

But the seeds of Davis's secessionist philosophy were crystal clear even as he insisted that he stood for the continuation of the United States: the founding ideals did not, to Davis, include any future movement toward ending slavery. Davis suggested that what bound revolutionary America together in opposition to the British was "community independence" and its assertion "against the powerful mother country"; he touted "the cause of State independence, and the right of every community to be the judge of its own domestic affairs." Such "community independence" and "State independence" meant that the national government must stay away from its inclinations toward curbing slavery. After all, said Davis, "Where is the grant of the Constitution which confers on the federal government a right to determine what shall be property?" Who, asked Davis, gave the federal government the "right to decide that [slavery] is a sin?"[1]

Davis wasn't merely referring to the federal government's power to disestablish slavery in slaveholding states. He was standing against the capacity of the federal government to regulate slaveholding in the federal territories, or to bar slaveholding in future states. According to Davis, then, our Americanism wasn't rooted in the principles of the Declaration of Independence, but in the pure contractualism of separate parties pursuing separate interests. America wasn't based on national principle. It was, in fact, based on power relations. The Constitution of the United States was merely a governing charter for a variety of separate interests, not a fulfillment of the promises of the Declaration of Independence.[2]

Davis's view was commonplace in the American South. Its most ardent articulator had been John C. Calhoun, the firebrand senator, secretary of state, secretary of war, and two-time vice president. Calhoun routinely disparaged the Declaration of Independence, infamously explaining on June 27, 1848, that its key proposition, that "all men are created equal," was patently false and superfluous: "[This proposition] was inserted in our Declaration of Independence without any necessity. It made no part of our justification in separating from the parent country." Calhoun described that key proposition at the heart of Americanism as a dangerous error, and lamented, "For a long time it lay dormant; but in the process of time it began to germinate, and produce its poisonous fruits." Those poisonous fruits: the American disinclination toward the continuation and spread of slavery.[3] The important part of the American founding was the Constitution alone—a contract between parties, which expressed few fundamental values.

If America was merely a system—a way of balancing competing political interests—then if the balance failed, so too would the system. In the view of the slaveholding states, that's precisely what happened. A separation was impelled.

Less than three years after touting the virtues of the Union at Faneuil Hall, Davis would resign his seat in the United States Senate—a decision he called the "saddest day of my life." Upon leaving the Senate, on January 21, 1861, Davis told the Senate that the Declaration of Independence supported his actions—its guarantees of liberty "ha[d] no reference to the slave." He declared "the right to withdraw from a Government which thus perverted threatens to be destructive of our rights," and threatened that should the North deny secession, "We will invoke the God of our fathers who delivered them from the power of the lion, to protect us from the ravages of the bear; and thus, putting our trust in God and in our own firm hearts and strong arms, we will vindicate the right as best we may."[4]

And so Jefferson Davis moved from an advocate of union to an advocate of separation. As he explained at his inauguration in Birmingham: the ties that bound Americans up with one another had dissolved, Davis stated, and "antagonisms are engendered which must and should result in separation."[5] A belief that only weak ties bound together Americans led to the dissolution of those ties. Davis formally became a secessionist.

The philosophy underlying secessionism never died. It merely metastasized. Today, Disintegrationists claim, just as Davis and Calhoun did, that America was based on a power arrangement, not on fundamental principle. Ironically, Disintegrationists make such claims on the basis of defending the same minority groups Davis and Calhoun targeted. But their larger point—that the entire American system is a hierarchy of power, not a system based on equally applicable principles—perversely reflects the secessionist view of American history.[6] In the Disintegrationist view, America is corrupt, a ruse and a sham promising liberty but actually guaranteeing tyranny. The American system, in this view, did not replace the Hobbesian war of all against all—it merely channeled that war into a system of dominance by white Americans, male Americans, straight Americans. All the high-minded talk about unalienable rights and delegated powers is simply kabuki theater. The true story of America is the story of the strong crushing the weak, both domestically and abroad.

If that premise is true, then America is not a true nation. America is, instead, an agglomeration of competing interests, forced together by circumstance and fate, damned to interminable struggle—or to divorce. America is the story of exploitation and greed, of patriarchy and abuse, of hierarchy and manipulation. The Declaration of Independence was a lie, not an unfulfilled promise; the Constitution was an enshrinement of slavery and brutality, not a bulwark against tyranny. America's story is an unending litany of horrors, punctuated by brief respites, always sliding back into the damnable bacterial soup from which we sprang. Progress is achieved in *spite* of America's nature, not because of it; deprivation and inequality and failure are natural results of Americanism.

Disintegrationist history relies upon three fundamental principles: first, that America was born in sin; second, that America has always reflected divided sects and hierarchies of power; and third, that America's role in the world has resulted in poverty, death, inequality, and injustice.

Now, as we have seen, this isn't true. America is unique; her philosophy, her culture, her history are indeed exceptional. But the goal of Disintegrationism is to dissolve the country's ties, not to reinforce them. Thus, to understand the Disintegrationist view of history, we must begin with a simple premise: for Disintegrationists, the purpose of history is *not* to understand the nature of America or even her development over time. The purpose of history is to provide fodder for the argument that modern problems are simply the latest manifestations of a deep-seated, incurable philosophic and cultural cancer lying at the heart of the United States. History is a weapon, not a bond. History isn't glue—it's acid. This viewpoint leads, inevitably, to disintegration. After all, if history is a constant source of division—if history is not only divisive but can never even be put behind us, if history is a bleeding artery, seeping sepsis into modernity—then the only answer is amputation.

The rise of Disintegrationist history reflects the binary thinking that so often attends our political debates. Instead of criticizing history that *has* sometimes been one-sidedly jingoistic, willing to paper over the ugly sins of America's past, Disintegrationists have posited that America herself is unfixable. America is either good or bad, say the

Disintegrationists. Since she cannot be good, she is bad. This dark and inaccurate view of the United States has become dominant in recent years. That's thanks to the generalized increase in Disintegrationist philosophy and culture, both of which suggest that individuals in America aren't responsible for their own successes or failures. Since America is, by all available metrics, incredibly tolerant and open, it has grown more and more difficult to find instances of true oppression in America. When demand for victimization narratives exceeded supply, Disintegrationists mine American history for such oppression, then declare that modern ills can be attributed to historical injustices. Sometimes, to some extent, that is the case. But the outsize attempt to pin modern, continuing ills on events centuries old wears thin for those who would rather solve problems than create them.

Unfortunately, as stated, there's a large market for the Disintegrationist narrative of victimization. It's quite difficult to debunk or fight the hovering miasma of "institutional" or "historic" discrimination—certainly far harder than debunking or fighting modern instances of such discrimination, which are routinely condemned by goodhearted people across the political spectrum. If you wish to suggest that systemic change is necessary in order to achieve an end goal, the most useful strategy is to blame the system. And if you can't blame today's system, simply label it an outgrowth of yesterday's.

The timing actually works both ways. Yes, Disintegrationists routinely blame the present on the past. But they also smear those they disagree with in today's politics by association with villains of the past. Once America's history has been painted as unremittingly bleak, it becomes a simple intellectual game to connect political views you don't like to negative historical antecedents. Conservative commitment to civil conversation rather than cancel culture becomes merely the latest manifestation of neo-Confederacy;[7] opposing gun control on slippery slope grounds becomes akin to opposing restrictions on slave trading in the nineteenth century.[8] Nearly every revisionist history book becomes merely the predicate for politicking against modern opponents.

As it turns out, Americans don't like being likened to the worst people in American history. And they're unlikely to want to share a country with those who continuously do so.

THE RISE OF DISINTEGRATIONIST HISTORY

Disintegrationist history truly began with the rise of American progressivism. In the attempt to rewrite the bargain between the American government and the American people, progressives had to retcon the philosophy of the American founding. In doing so, they relied on historicism—a philosophy suggesting that all institutions and ideas are products of time and place, and thus completely fungible. Historicism cut directly against the notion of eternal, unchanging, inalienable natural rights—so it was a solid foundation for historical revisionism about the American founding. Thus, Woodrow Wilson, a devotee of historicism, stated that the Declaration of Independence "expressly leaves to each generation of men the determination of what they will do with their lives, what they will prefer as the form and object of their liberty, in what they will seek their happiness. . . . The ideals of liberty cannot be fixed from generation to generation; only its conception can be, the large image of what it is. Liberty fixed in unalterable law would be no liberty at all."[9]

To do away with the founding conception of rights and government, then, would require the rewriting of history. That rewriting would need to be justified, in turn, by a shift in the philosophy of history: instead of focusing on the *reality* of the past, it would focus on the past *through the lens of the present*. Frederick Jackson Turner, president of the American Historical Association, stated in 1910 that history ought to be separated from objective study, and ought to instead be activated in service of "points of view furnished by new conditions which reveal the influence and significance of forces not adequately known by the historians of the previous generation." Carl Becker, a deeply influential progressive historian, suggested as much: "To me, nothing can be duller than historical facts, and nothing more interesting than the service they can be made to render in the effort to solve the everlasting riddle of human existence." Becker would later explain, "in truth, the historical fact is a thing wonderfully elusive after all, very difficult to fix, almost impossible to distinguish from 'theory,' to which it is commonly supposed to be so completely antithetical." The purpose of history was to "convey an intelligible meaning of what happened."[10]

Becker clarified in 1931, in a famous lecture to the American Historical Association, "The appropriate trick for any age is not a malicious invention designed to take anyone in, but an unconscious and necessary effort on the part of 'society' to understand what it is doing in the light of what it has done and what it hopes to do."[11] In light of what society *hopes to do*—that phrase represents a complete reversal of the time direction of history from looking backward to looking forward. Fellow New Historian James Harvey Robinson summed up the goals of the new orientation: "Often history will work where nothing else will. It very gently modifies one's attitude."[12]

This philosophy—dubbed New History—would have an immediate impact on the American view of history. It provided the grounds upon which historical revisionists—particularly radical revisionists—could build. If America had to move beyond its roots, it would be necessary to rewrite those roots into something deeply embarrassing. The easiest way to do that would be selectively ignoring evidence in favor of a particular thesis.

The earliest prominent historian to perform that function was Charles Beard, author of *An Economic Interpretation of the Constitution of the United States* (1913). Beard, a professor at Columbia University, suggested that the founding fathers had put together the Constitution of the United States in order to protect their own economic interests. Beard described the founders as "a small and active group of men immediately interested through their personal possessions in the outcome of their labors," as opposed to the "propertyless masses," who were "excluded at the outset from participation." The founders weren't idealists, but oligarchs. The Constitution, Beard concluded, was "essentially an economic document." Beard's "discovery" made him famous—by 1952, his books had sold over 11 million copies.[13] Beard would also argue, along with his wife, Mary, that American business enterprise in the Gilded Age was an unending story of exploitation, rather than, in the main, a story of unprecedented American economic growth and progress, enabling better living for tens of millions. Unsurprisingly, Beard was a committed critic of capitalism, and a deep believer in the perfectibility of man; he wrote that "physical and social health must be made the basis of education."[14] It was no coincidence that Robinson and Beard would, together with Herbert Croly, found the New School for Social Research—an institution driven by the blurring of boundaries between progressive world-shaping and "scientific" research.

The economic take on history would infuse future history teaching. And while it turned out that Beard's work was seriously flawed—historian Forrest McDonald duplicated Beard's research and found "no correlation" between the founders' "economic interests and their votes on issues in general or on key economic issues"[15]—his historicism would set the precedent for future class-based renderings of American history. Staughton Lynd, a neo-Marxist who agreed broadly with Beard's class-based rewriting of the American founding, would pen *Intellectual Origins of American Radicalism*, in which Lynd invested the founding's promises of liberty and the development of

the abolitionist movement with socialistic undertones; Lynd acknowledged that he was "more and more committed to the thesis that the
professor of history should also be a historical protagonist."[16] Richard
Hofstadter, who cited Beard as his chief intellectual influence, became
the most praised historian of midcentury; he started his political career
as a committed communist (he actually joined the Communist Party
in 1938), but ended it as a two-time Pulitzer Prize winner. Yes, he de-
spised capitalism—he openly declared that hatred[17]—but he was able
to channel that dislike into a high-minded critique of everything from
American industry to American conservatism. George Will rightly
characterized Hofstadter's generalized outlook as scorn for American
conservatism: "Hofstadter dismissed conservatives as victims of character flaws and psychological disorders—a 'paranoid style' of politics
rooted in 'status anxiety,' etc. Conservatism rose on a tide of votes cast
by people irritated by the liberalism of condescension."[18]

The most popular of the class-based revisionists, however, was
clearly Howard Zinn. Zinn's poorly sourced, extraordinarily biased
history of America, *A People's History of the United States*—in which
America is responsible for nearly every ill of the past two centuries—
had sold more than two million copies at the time of his death.[19]
Zinn was an ardent defender of subjectivism in history. And his subjective point of view was purely Marxist—according to Zinn, who
spent nearly his entire adult life in flirtation with open communism,
the Marxist nostrum that history itself is merely class struggle was
"undeniably true, verifiable in any reading of history."[20] But Zinn's
view wasn't merely Marxist—it was just as fully and openly Disintegrationist. In the opening pages of *A People's History*, Zinn spills his
agenda: to rewrite history in the name of the people, so as to explode
"pretense . . . that there really is such a thing as 'the United States,'
subject to occasional conflicts and quarrels, but fundamentally a community of people with common interests." In Zinn's view, there is no

"'national interest' represented in the Constitution, in territorial expansion, in the laws passed by Congress, the decisions of the courts, the development of capitalism, the culture of education and the mass media." All of history is merely power hierarchy. The United States is merely a competition "between conquerors and conquered, masters and slaves, capitalists and workers, dominators and dominated in race and sex." And, says Zinn, "in such a world of conflict, a world of victims and executioners, it is the job of thinking people, as Albert Camus suggested, not to be on the side of the executioners."[21]

Who are those executioners? The long-standing heroes of American history, who are *actually* villains. To that end, Columbus became a brutal warlord victimizing peaceful natives; the founding fathers tyrants simply replacing one tyranny with another, papering over that tyranny with the creation of "a nation, a symbol, a legal unity called the United States," designed to "create a consensus of popular support for the rule of a new, privileged leadership";[22] Abraham Lincoln a mere political operator rather than emancipator and the Civil War a battle by the American government "not to end slavery, but to retain the enormous national territory and market and resources";[23] the men and women who waged World War II, representatives of a racist tyranny and the ultimate human rights violator, winning victory on behalf of "militarism, racism, imperialism," which would be absorbed into "the already poisoned bones of the victors";[24] anticommunists, representatives of imperialism and militarization and suppression of dissent at home; and communists, heroes, particularly the Viet Cong.[25] In a book of nearly seven hundred pages, with no footnotes, Zinn somehow manages to avoid anything worthwhile or decent about the greatest country in the history of mankind.

Zinn's history is shabby, his politics downright despicable. Yet precisely because Zinn's history is so overtly anti-American, it has achieved the status of canon among those who prefer to view themselves as "speaking truth to power." Sam Wineburg of Stanford University explains, "*A People's History* speaks directly to our inner Holden Caulfield. Our heroes are shameless frauds, our parents and teachers conniving liars, our textbooks propagandistic slop. They're all phonies is a message that never goes out of style."[26] Zinn's disciples are legion. They range from Matt Damon and Ben Affleck—who famously had Will Hunting name-check Zinn in an intellectual conversation with a Harvard professor, who also talked up radical anti-American revisionist Noam Chomsky—to A. J. Soprano in *The Sopranos*. They include Pulitzer Prize winners like Eric Foner.[27] Professors across the country assign Zinn's *People's History of the United States* in their classrooms; increasingly, high schools use curricula built around Zinn's writings. Wineburg says that "in many circles, it has *become* the dominant narrative. The book appears on university reading lists in economics, political science, anthropology, cultural studies, women's studies, ethnic studies, Chicano studies, and African American studies, in addition to history."

The rewriting of American history wasn't merely revised through the prism of class dynamics; it was quickly revised through the prisms of race, gender, and sexual orientation. Critical theory, as posited by the Marxist-influenced Frankfurt School, sought to reexamine American history as a story of hierarchical dominance by one powerful group—white men—over all other groups. This view casts America not as a nation but as a series of competing interest groups, which can remake America only by overthrowing the corrupt status quo. As applied to race, critical theory suggests—as did Zinn's class struggle theory—Disintegrationism. The core of critical race theory (CRT) lies in the propositions that racism is ordinary and permanent; that whiteness and property coincide; that history is told only by dominant groups, and requires a counternarrative; and that color-blindness is a myth, and that the notion of equality of rights is itself a reflection of color hierarchy.[28] The net result is that the system itself must be torn down, and that all the promises of the Declaration of Independence and Constitution are lies. As Derrick Bell, a founder of critical race theory, wrote, "the concept of individual rights, unconnected to property rights, was totally foreign" to the founding fathers, and thus, "despite two decades of civil rights gains, most Blacks remain disadvantaged and deprived because of their race."[29] Victimization can never be alleviated under the current system. Only the disintegration of that system will rectify the modern imbalances that simply mirror the sins of the past.

THE DOMINANCE OF DISINTEGRATIONIST HISTORY

The perversion of American history has become the mainstream narrative on the left side of the political spectrum—and since the political left dominates the educational system, Disintegrationist history has become establishment history. That historical revisionism has reached the highest levels of American politics. In March 2008, for example, Barack Obama attempted to push back against accusations that he had touted Reverend Jeremiah Wright, a vicious race-baiter and anti-Semite. He did so by blaming American history writ large, justifying Wright's radicalism as a reflection of the "complexities of race in this country," and suggesting, "Understanding this reality requires a reminder of how we arrived at this point." The problem, as it turned out, wasn't the despicable ideology promoted by Wright—it was American history.[30] This was a constant theme of Obama's presidency. In 2015, Barack Obama, the president elected by 69.5 million Americans in 2008, explained that all good in America had grown from poisonous seeds: "The legacy of slavery, Jim Crow, discrimination in almost every institution of our lives. You know, that casts a long shadow and that's still part of our DNA that's passed on."[31] America could only move beyond that history through "fundamentally transforming the United States of America," said Obama; his wife, Michelle, explained, "we're going to have to change our traditions, our history; we're going to have to move into a different place as a nation."[32]

Obama's mantle was picked up after his presidency by the thought-leaders of the *New York Times*, who declared in 2019 that America's true founding had taken place *not* in 1776, but in 1619.[33] As Adam Serwer correctly diagnosed at *The Atlantic*, "The most radical thread in the 1619 Project is not its contention that slavery's legacy continues to shape American institutions; it's the authors' pessimism that a majority of white people will abandon racism and work with black Americans toward a more perfect union. Every essay tracing racial injustice from slavery to the present day speaks to the endurance of racial caste."[34]

Pulitzer Prize–winning Revolutionary War historian Gordon Wood called it "so wrong in so many ways," and fellow Pulitzer Prize–winning Civil War historian James McPherson called it an "unbalanced, one-sided account" that "left most of the history out."[35] The newspaper put together a curriculum in conjunction with the Pulitzer Center designed to be taught to schoolchildren; more than one thousand teachers have taken up the offer.[36]

In response to the claim that Disintegrationist history tears apart the nation—and is selective in its focus on events, to the exclusion of broader American exceptionalism—Disintegrationists flip the script. *They* are not the causes of lack of unity—traditional advocates for Unionist history are. Those Unionists have marginalized the voices of those silenced for too long. To teach traditional history is to participate in oppression. Traditional history must be discarded in favor of fragmentary approaches to history, as refracted through the prism of race; in the words of Professor Frances Negrón-Muntaner of Columbia, ethnic studies courses are designed to "upend colonial (including white supremacist) epistemologies, institutions, and power structures; and . . . generate decolonial narratives, subjectives, and forms of organization."[37] Teaching traditional Western civilization means elevating it; thus Western civilization must not be taught any longer. And it isn't: as of 2010, not a single top university required students to take a course in Western civilization.[38]

Increasingly, Americans don't know history at all. And why should they? To understand America's history properly would be to learn about the past—and learning about that past does not achieve the purposes of the Disintegrationists. Better that history should be taught through the prism of politics than that Americans should learn a fully shaded history of the country. The rise of class-based and race-based theories of American history has been accompanied, unsurprisingly, by a skyrocketing ignorance of American history altogether. On college campuses, colleges have shifted their focus toward "diversity requirements"—classes that, at UCLA for example, "substantially address[] racial, ethnic, gender, socioeconomic, sexual orientation, religious or other types of diversity."

These are the sorts of classes in which, according to *The Atlantic*, "Students talk about Trayvon Martin in the same breath as slavery and the convict lease system."[39] Across the country, for decades, social studies has supplanted history in high schools and middle schools—an attempt to make history "relevant" by replacing history teaching with indoctrination into a particular brand of politics. Education historian Diane Ravitch lamented the decline of American history teaching back in 1985 in the pages of the *New York Times*, acknowledging that history had been "absorbed" into politically driven "social science."[40]

It's not merely that Disintegrationists offer an alternative version of history—that would, at least, allow for the presentation of a more traditional American history. It's that Disintegrationists castigate those who teach traditional American history as blinkered, foolish, backward—and insist that such teaching be relegated to the supposedly ethnocentric, patriarchal past. Nikole Hannah-Jones, matriarch of the 1619 Project, sarcastically tweeted upon hearing of McPherson's criticism, "LOL. Right, because white historians have produced truly objective history," and went on to disparage his credentials as a historian.[41]

Michael Harriot of *The Root*, a self-described "world-famous wypipologist" (white people-ologist), similarly ripped into Peter Kirsanow of *National Review* for pointing out that the 1619 Project erased white contributions to the country. In a piece titled "Black History, According to White People," Harriot explained that the traditional view of American history was "at best, a mythical rendition of the past retold by white mouths so long that it has become the truth. It whitewashes our martyrs and obscures the ignominious white supremacy that built this country. It diminishes the truth of the slaveholding Founding Fathers and paints this country as a beacon of liberty and justice." Kirsanow—a "motherf***er," he said—wrote "whitely," and studied history "as white people are wont to do."[42] It was an unfortunate fact for Harriot's viciously racialized narrative that Kirsanow happens to be black.

But this reductionist view of the study of history owes much to the critical studies theories that became dominant decades ago. The view also happens to be blatantly racist: judging a historian's work by his skin color should have gone out of style long ago. But that blatant racism is a *required* element of critical race theory, just as a focus on "economic privilege" is a requirement of class-based theory. American history must be rewritten by the oppressed and the marginalized, because American history is the story of oppression and marginalization.

The purpose of this historical rewriting isn't reparation—it's not about fixing the problem or making absolution for America's sins so that we can move forward together. That would be the role of traditional American history—acknowledging our flaws along with our qualities, but understanding that America's philosophy, culture, and history are rooted in glory rather than evil. In the Disintegrationist view, though, America is inherently unfixable. Ta-Nehisi Coates, while making the case for slavery reparations, indicts the entire American system: "the crime with which reparations activists charge the country implicates more than just a few towns or corporations. The crime indicts the American people themselves, at every level, and in nearly every configuration." Coates adds, "Perhaps after a serious discussion and debate . . . we may find that the country can never fully repay African Americans." After indicting the American people themselves, at every level, in every configuration, does anyone truly believe Coates's "perhaps"?[43]

BORN IN SIN

As we have noted, Disintegrationist history teaches three fundamental principles: first, that America was founded in evil; second, that America is irredeemably divided and can never escape her past absent dismantling her founding principles; third, that America has been, on net, terrible for her citizens and terrible for the world.

This third principle, in particular, is provably false, as we have discussed. But Disintegrationists simply assume the third principle by virtue of the first two: America was born in sin, and can never be redeemed. Essentially, Disintegrationists claim that America, which has dragged the world out of ultimate impoverishment, cannot be supported on any *moral* level thanks to America's sins. This claim makes a simple moral appeal: America may have generated great wealth and prosperity and peace, but that's in *spite* of its founding principles, not because of them. The utilitarian greatness of America simply cannot answer for America's historical and current evil. America's true founding can be found in its treatment of natives, centuries-long perpetuation of barbarous slavery, industrial exploitation of workers, and corruption by certain industrial and financial barons. America was born in sin.

Disintegrationists, therefore, date America's founding not to 1776, or even to Jamestown or Plymouth Rock, but to two dates: 1492, and 1619. The first signifies the West's arrival in the New World, bringing with it plague, imperialism, exploitation, and death; 1619 signifies Americans' reliance on chattel slavery to build its wealth.

To push the narrative that America is deeply exploitative, Disintegrationists focus in on the fact that Europeans who first arrived on the shores of the New World weren't arriving on empty land: they came and dispossessed and mistreated the natives, who were peacefully pursuing their lives before the advent of Western colonialism. Why, exactly, should this be particularly relevant to modern politics? According to Disintegrationists, the dispossession of Native Americans underscores the brutality of Western civilization, and more broadly of free markets.

In this view, European property rights *caused* dispossession. In this narrative, Christopher Columbus comes in for particular ire; indeed, Howard Zinn opens his masterwork by contrasting the Arawaks of the Bahamas Islands, who he says were "very much like Indians on the mainland, who were remarkable (European observers were to say again and again) for their hospitality, their belief in sharing," with the evil Europeans, who were "dominated as it was by the religion of popes, the government of kings, the frenzy for money that marked Western civilization and its first messenger to the Americas, Christopher Columbus." In this narrative, Columbus was merely Western brutality writ small, seeking to enslave the peaceful natives.[44] And the natives, by contrast, were the ideal of humanity: "Columbus and his successors were not coming into an empty wilderness, but into a world . . . where human relations were more egalitarian than in Europe, and where the relations among men, women, children, and nature were more beautifully worked out than perhaps any place in the world."[45] And this narrative has come to dominate the teaching of American history.

Needless to say, even were this one-sided history true—and Columbus's personal treatment of the natives was exaggerated in extraordinary fashion by Zinn through selective quotation[46]—this doesn't debunk the free-market principles that have animated America any more than pointing out that muggers victimize old ladies from whom they steal purses. Yes, that is an example of dispossession; no, it is not an example of capitalistic brutality. And even should the mugger go on to invest his stolen gains in the stock market and become a millionaire, this would not make the stock market exploitative, but the mugger. But according to Zinn, the West's reliance on property rights not only corrupted the New World, it then infused its free-market brutality throughout American history.

More important, Zinn's argument that America was born in sin and thus is irredeemable absent a complete abandonment of its European, property-bound roots is idiocy of the highest degree. European culture was not uniquely violent, nor was expansionism a uniquely European invention. One need not minimize the brutality of Europeans toward Native Americans in order to recognize this. Zinn's Rousseauian view of pre-European American life simply is not supported by the archaeological record. As Harvard's Steven Pinker rightly states, "Contra leftist anthropologists who celebrate the noble savage, quantitative body-counts—such as the proportion of prehistoric skeletons with axemarks and embedded arrowheads or the proportion of men in a contemporary foraging tribe who die at the hands of other men—suggest that pre-state societies were far more violent than our own."[47] This is important, because context provides knowledge: yes, the European encroachments on the American continent were bloody and awful, but bloody and awful conflict was nothing new on the continent itself. Claims of genocide by Columbus are wildly wrong; the depopulation of natives after contact with Europeans was almost entirely the result of disease to which natives had no immunity.

By singling out European society as uniquely evil and relying on the myth of peaceful native socialism, the goal is to debunk Western thought as pathologically incapable of good: whatever strengthens the West strengthens its evil. But again, human aggression preceded the actual American founding—with its glorious ideas of individual rights, granted by God, preexisting government—by hundreds of thousands of years. That true founding idea has *reduced* colonialism and imperialism over time, which is why the United States may be the most powerful country in the history of the world, but does not rule the largest territorial empire in the history of the world. Traditional unionist historians do not need to ignore the Trail of Tears or the enslavement of the Arawaks in order to cover for the West. And, as we've seen, they don't. But Disintegrationists must ignore both the contemporaneous and subsequent history of the world in order to cover for their own ahistorical perspective.

The second American founding, according to the Disintegrationists, came in 1619, with the importation of the first African slave into the new colonies. This line has been most famously promulgated in our day by the 1619 Project, which declared America's actual founding date the importation of the first African slave to the continent. The *Times* stated that 1619 was America's "true birth year," that "Out of slavery—and the anti-black racism it required—grew nearly everything that has truly made America exceptional: its economic might, its industrial power, its electoral system, its diet and popular music, the inequities of its public health and education, its astonishing penchant for violence, its income inequality, the example it sets for the world as a land of freedom and equality, its slang, its legal system and the endemic racial fears and hatreds that continue to plague it to this day." The newspaper of record declared that it aimed to "reframe the country's history" by placing slavery at its center.[48] The goal here is obvious: to suggest that America's values are inherently white supremacist and tyrannical, rather than rooted in liberty.

This would all simply be interesting, if it hadn't been created with a divisive political agenda in mind: to suggest that America's birth took place in sin, and that every subsequent sin of America can be traced to those original sins. Fourteen ninety-two finds its echoes in the Mexican-American War, in the Philippines, in Hawaii, in Vietnam, in Iraq; 1619 finds its echoes in railroad tycoons utilizing immigrant labor, monstrous corporations outsourcing jobs to China and Mexico, higher imprisonment rates for black Americans. Seventeen seventy-six, in this view, fades into the distance; it was a mere mask, which when removed reveals the evil face of colonialism, racism, and exploitation. Thus the Disintegrationists portray America's founding as *truly* a reflection of 1492 and 1619, an attempt to reinforce class and race dominance. The Declaration of Independence was a lie, the Constitution a document enshrining oppression.

The economic narrative—that the Declaration of Independence was written by those who sought to perpetuate class dominance—is blatantly untrue. Not only did the founding fathers often vote without regard for their own property interests,[49] but America quickly became the land of opportunity for citizens treated properly under the Constitution. Alexis de Tocqueville, as we have seen, paid tribute to both the entrepreneurial spirit of Americans and to the ability of American citizens to pursue their economic interests. "Almost all Americans are comfortable," Tocqueville observed. "In America most of the rich have begun by being poor. . . . Men show themselves to be more equal in their fortunes and in their intelligence or, in other terms, more equally strong than they are in any country in the world and than they have been in any century of which history keeps a memory."[50] He added, "The poor of America, compared to those of Europe, could often appear to be the wealthy."[51]

The racial narrative takes a germ of truth—that America did, in despicable fashion, tolerate and expand slavery and white supremacy before the vitiation of the former through the Civil War and the latter through decades of civil rights work—and builds an entire narrative of falsehood upon it. Disintegrationists argue that racism was ineradicably and unalterably baked into the American cake: Michelle Alexander, author of *The New Jim Crow*, for example, recently wrote in the *New York Times* that our Constitution was "defined" by "white supremacy." And she adds that such white supremacy has lasted the test of time—that all evidence to the contrary, white supremacy has merely shapeshifted rather than having been largely eviscerated:

> The politics of white supremacy, which defined our original constitution, have continued unabated—repeatedly and predictably engendering new systems of racial and social control. Just a few decades ago, politicians vowed to build more prison walls. Today, they promise border walls. The political strategy of divide, demonize and conquer has worked for centuries in the United States— since the days of slavery—to keep poor and working people angry at (and fearful of) one another rather than uniting to challenge unjust political and economic systems. At times, the tactics of white supremacy have led to open warfare. Other times, the divisions and conflicts are less visible, lurking beneath the surface.[52]

"Divide, demonize, and conquer" can be directed both ways. To argue that the Constitution was designed *on behalf of white supremacy* is to ignore both the context and content of the document. That would be like arguing that the New Deal was performed on behalf of sexism: the cultural backdrop of a country, as ugly as it may be, is merely the backdrop, not the motivating ideal. Along the same lines, Jim Crow, which lasted a full century after the Civil War, was an overt, violent, white supremacist violation of the Constitution—and stood in stark contrast to our founding ideals, as black leaders from Booker T. Washington to Martin Luther King Jr. pointed out.

Alexander's argument most obviously downplays racial progress. And herein lies the *true* goal of Disintegrationist history: to suggest a continuum between America's birth in sin and her current ills.

DAMNED FOR ETERNITY

In the Disintegrationist view, every ill of American society can be traced to America's monstrous beginnings; every sin of American society is evidence of America's malformed origins. In fact, even America's *greatnesses* were poisoned at the outset—all points in America's favor thus become strikes against her. Because of America's long history of racism, the chief arrows in the Disintegrationist quiver, aimed directly at the goodness of America, are slavery and Jim Crow. And these arrows are used for any and every possible purpose.

Thus, for example, Matthew Desmond, a Princeton sociology professor writing for the 1619 Project, argued that America's capitalism— the most successful engine for wealth in the history of the world—is both brutal and was formed by the culture of slavery. In his article "In Order to Understand the Brutality of American Capitalism, You Have to Start at the Plantation," Desmond actually makes the stunning case that oversight of slaves somehow provided the predicate for companies that implement workplace supervision efforts to chart employee time.[53] Not only is the claim patently silly on its face, but it is also completely ignorant of American economic history, which transitioned to industrial capitalism—as against the agrarian slave-based economy of the South—after the Civil War.

The less foolish version of the argument is that America's wealth was built not by capitalism, but on the back of slavery. Some advocates of this argument push for slavery reparations, explaining that all non-black Americans benefited from slavery. Nearly all argue that color-blind regimes of individual rights ought to be overturned in favor of legal arrangements to favor blacks, including affirmative action and government-provided benefits programs.

These suggestions rest on two *other* claims: first, that American citizens broadly benefited from slavery and Jim Crow; second, that today's inequalities are due to the legacy of slavery. Each claim is untrue in the main.

The claim that American citizens benefited broadly from slavery does not hold water. America's wealth was not dependent on slavery; slavery destroyed millions of lives, and created wealth for a select few southern aristocrats, but left the South extraordinarily economically underdeveloped overall. Slavery reparations from slaveholders to slaves would have been fully justified; the lack of such reparations is a great horror and tragedy. But to argue that the grandson of a Lithuanian immigrant ought to pay the great-great-great-great-granddaughter of a freed slave *today*, or that Colin Powell's child ought to receive reparations from a slain Union soldier's great-great-great-great-grandson—that argument rests on faulty premises.

Slavery was indeed an important part of the American economy. It was also a *backward* part of the American economy. Freeman labor is simply more economically efficient than slave labor. It's no wonder that Alexis de Tocqueville, in *Democracy in America*, described the South as a society gone to "sleep," where work was "degraded," an area filled with "idle" men.[54] In fact, the growth of the southern economy remained slow throughout the Jim Crow period—black labor fled to the North. Only as Jim Crow waned did growth rates in the South suddenly spike—leading, not surprisingly, to a population movement *back into those states*.[55] If slavery had been an economic winner, the South wouldn't have been roundly defeated by the industrialized North. The end of slavery meant a drop in agricultural capital in the United States, but a massive uptick in industrial and housing capital, as well as other domestic capital. Our labor stock soared, and our capital stock soared in the aftermath of the Civil War. This experience isn't shocking. As Scott Sumner of the Library of Economics and Liberty points out, "Brazil didn't abolish slavery until the 1880s, and did worse than America. It also did worse than countries to the south of Brazil." Free alienation of labor drives prosperity, which is why free-market countries are by far the most prosperous.[56] Arguing that Americans broadly benefited from slavery is like arguing that Europeans broadly benefited from feudalism.

What about Jim Crow? Again, Jim Crow benefited a select few at the expense of the broader American economy. Jim Crow prevented competition, led to a mass migration of blacks from the South to more tolerant parts of the country, and generally depressed the southern economy. Had the economic power of black America been unleashed, the American economy would have *exploded*.

But it is certainly true that Jim Crow had a lasting impact on the racial wealth gap. Economic history always has an effect on the economic present. Certainly policies like redlining, while outlawed by the federal government in 1977 under the Community Investment Act, prevented black Americans from investing in housing at the same rates as whites; segregation and discrimination prevented black businesses from gaining access to capital at the same rates as whites, and led to underfunding of black educational institutions. All of that is undeniable.

How much of the racial wealth gap in the United States *today* is explained by such historic discrimination? And how much have America's sins in this regard been alleviated by subsequent changes of law and heart? In the view of Disintegrationists, not in the slightest. In the view of reality—an enormous amount.

In order to compare the impact of past discrimination on future wealth creation, we must look to the performance of control groups— nonblack minorities once discriminated against in law. Coleman Hughes of Quillette points to the performance of Japanese-Americans, who were banned in fourteen states from owning land from 1913 to 1952, and experienced mass internment during World War II. One analysis found that Japanese-Americans' median wealth will soon surpass that of white Americans. That's thanks to Japanese-American earning power: as Hughes points out, "by 1970 census data showed Japanese-Americans out-earning Anglo-Americans, Irish-Americans, German-Americans, Italian-Americans, and Polish-Americans." Similarly, Jewish Americans had a 7-to-1 household wealth advantage over conservative Protestants as of 2003.[57]

No group in America has been treated as horrifically over time as black Americans. That is undoubted. But it is true for every group that wealth gaps can only be overcome through increases in *income*. In order to rectify the wealth gap over time, then, the true focus *should* be on the income gap. And here the historical explanation for the white-black gap utterly falls apart.

When it comes to income, disparities in modern America are due, in the main, to individual rather than structural issues. That is why Asians now out-earn whites in America by a wide margin; it is why Hispanic Americans are moving up in the income distribution across generations, while black Americans are not, according to researchers from Stanford, Harvard, and the U.S. Census Bureau. That study actually found that "growing up in a high-income family provides no insulation from these disparities. . . . Black children born to parents in the top income quintile are almost as likely to fall to the bottom quintile as they are to remain in the top quintile." Furthermore, that same study found that the black-white income gap was driven "entirely" by differences in men's, not women's, outcome. In other words, black women "earn slightly *more* than white women conditional on parent income." The study found "little or no gap in wage rates or hours of work between black and white women."[58]

It is nearly impossible, then, to attribute current income gaps directly to racism. It is worth noting that black immigrants to the United States have outperformed the descendants of slaves economically. Forty years ago, Thomas Sowell observed, "West Indians in the United States have continued to hold sizable advantages over American Negroes in incomes and occupations."[59] It's still true today.

What, then, are the easily identifiable factors that lead to lower income mobility for black Americans? First, lack of fathers in the home: single motherhood is one of the most powerful predictors of intergenerational poverty. Approximately seven in ten black children in America grow up without a father in the home.[60] And as Barack Obama acknowledged in 2008, too many black fathers "have abandoned their responsibilities, acting like boys instead of men. We know the statistics—that children who grow up without a father are five times more likely to live in poverty and commit crime; nine times more likely to drop out of schools and 20 times more likely to end up in prison. They are more likely to have behavioral problems, or run away from home or become teenage parents themselves. And the foundations of our community are weaker because of it."[61] The increase in single motherhood in the black community has nothing to do with either Jim Crow or slavery; in 1963, just 24.2 percent of births in the black community were out of wedlock.[62]

Other factors include higher educational dropout rates (8.0 percent of black male students drop out of school, compared with 4.9 percent of white males);[63] arrest rates (while the rate of black male imprisonment has dropped dramatically over the past fifteen years, 2,613 black males per 100,000 black males were in prison as of 2015, compared with 457 white males in prison per 100,000 white males);[64] and spending and saving patterns.[65]

The Disintegrationist retort is to blame each of these factors, in turn, on institutional racism. That makes it easy to avoid explaining exactly what is causing these problems. Single motherhood was far lower before the Civil Rights Act than it is today; crime rates spiked *after* the end of Jim Crow; and educational rates have stagnated for years. When doctors don't know what is causing a disease or how to cure it, they call it a syndrome. When sociologists don't know what is causing a problem or how to cure it, they call it systemic.

The difference, of course, is that sociologists often *do* know what is causing the problem. But encouraging better individual decision making doesn't fit within the Disintegrationist worldview. Disintegrationist history is designed to explain everything, but it fixes nothing.

And that, of course, is the point.

CONCLUSION

The third and final step in destroying America is convincing citizens that America represents fruit of the poisonous tree: that America was founded in evil, and that there is no arc to history. Rather than a story of ever-expanding freedoms and prosperity rooted in the true seed of eternal and good values, America become a story of internecine warfare and brutality, with only the victims changing.

During the 2020 election cycle, Beto O'Rourke, a short-lived Democratic candidate for president, summed up his view of American history. "Racism in America is endemic," O'Rourke railed. "It is foundational. We can mark the creation of this country not at the Fourth of July, 1776, but August 20th, 1619, when the first kidnapped African was brought to this country against his will and in bondage and as a slave, built the greatness and the success and the wealth that neither he nor his descendants would ever be able to fully participate in and enjoy." He then cited this egregious history as the rationale for everything from disparities in black-white maternal mortality statistics to the racial wealth gap.[66] This was obviously pandering by the newly woke O'Rourke, whose precipitous drop in the polls soon forced him from the race.

But the fact that O'Rourke had to pander by deliberately misinterpreting American history should have been troubling to observers of any stripe. And O'Rourke was hardly the only Democratic candidate to do so. Senator Cory Booker of New Jersey exclaimed, "We have systemic racism that is eroding our nation from health care to the criminal justice system. And it's nice to go back to slavery, but dear God, we have a criminal justice system that is so racially biased, we have more African Americans under criminal supervision today than all the slaves in 1850." Mayor Pete Buttigieg of South Bend, Indiana, similarly blamed all disparities on America's history: "We know that the generational theft of the descendants of slaves is a part of why everything from housing to education to health to employment basically puts us in two different countries."[67] His ignorance of history didn't begin during his presidential run, either; in 2014, Buttigieg claimed, "the people who wrote the Constitution did not understand that slavery was a bad thing."[68] Former vice president Joe Biden blithely told an unbothered audience that "English jurisprudential culture, a white man's culture" was responsible for the mistreatment of women in the United States, and added vociferously, "It's got to change, it's got to change."[69]

If America wishes to be great, that greatness will require coming to grips with her history—its evils yes, but its glories, too. That means repenting our sins, but it also means celebrating our virtues and our victories. It means rejecting the simplistic anti-American history of the Disintegrationists and embracing the full Unionist history of the United States. If we do not unify around our common past, we can have no common future.

CONCLUSION

In 2016, President Trump campaigned on a simple slogan: "Make America Great Again." It wasn't clear just what Trump meant by the slogan—Trump advocates will say he speaks in pithy aphorisms, Trump opponents that he speaks in foolish bumper stickers—but the slogan proved popular nonetheless. That's because, in their hearts, Americans *do* believe that America was *always* great. We didn't always live up to our greatness—but we were always great. Proof of that greatness could be found in the Declaration of Independence, the Constitution of the United States, the Northwest Ordinance, the writings of Jefferson and Adams and Madison and Washington, the Monroe Doctrine, the Civil War, the Gilded Age, World War I, World War II, the Cold War, and the war on terrorism; from the defeat of Nazis and communism to putting a man on the moon, from the agrarian beginnings of the United States through industrial revolution and on into the informational age, America was great.

How could Americans think that? Because to most Americans, the story of America is a story of her extraordinary and unparalleled founding *philosophy* and her uniquely thriving and chaotic *culture*—and their ideals manifesting with greater and greater frequency in her *history*. To most Americans, America is about, and was always about, the Declaration's beliefs in natural law rooted in the human ability to reason, in equal rights under law, in individual rights preexisting government, in a government whose legitimacy sprang from the consent of the people and in its own protection of those rights—and in constitutional structures designed to implement those beliefs. To most Americans, America is about, and was always about, rights to religious freedom, to freedom of speech, to freedom of self-defense, to economic freedom—and about the duties provided and enforced by thriving social institutions inculcating virtue. To most Americans, America is about, and was always about, the gradual realization of those beautiful promises, rights, and institutions for all Americans.

Perhaps most of us never thought about all of that. Perhaps we just saw the flag flying and *felt* it, or watched a Fourth of July parade and *knew* it. Perhaps we never thought enough about what the flag stood for, or what exactly our founders risked their lives on behalf of when they signed the Declaration of Independence. Perhaps many of us never thought about the downside of America, too—all the times that we didn't live up to our promises, our Declaration, our Constitution, our churches and our neighbors, and our flag. That is our fault. But that is not America's fault. America was always great, even if she was never perfect. America has spent more than two centuries moving toward a more perfect union—but that union was founded on eternally good and true principles, planted amid a culture of rights and duties, and eventually sprouting into the greatest nation in the history of mankind.

Nonetheless, "Make America Great Again" became highly controversial. Part of that was due to the fact that Trump is the King Midas of controversy: everything he touches turns to chaos. And certainly Trump's inability to explain the meaning of the slogan—along with his own vulgarity, his ignorance of American history and philosophy, and his less-than-stellar adherence to cultural foundations of American rights—left room for Trump's opponents to misinterpret the slogan.

But Trump's most ardent opponents weren't content to point any of that out, or to ask Trump for clarification of just what he meant. They weren't even content to leap to the conclusion that Trump's sepia-toned nostalgia was for a *worse* time—a time of segregation or sexism or imperialism. Instead they leapt to a photo-negative slogan, a slogan that provided the most stark contrast of all: "America Was Never Great."

Protesters chanted that slogan. And that slogan became de rigeur for many intellectuals on the political left. Governor Andrew Cuomo of New York stated, "We are not going to Make America Great Again. It was never that great. We have not reached greatness. We will reach greatness when every American is fully engaged. We will reach greatness when discrimination and stereotyping against women, 51 percent of the population, is gone, and every woman's full potential is realized and unleashed."[1] Obama attorney general Eric Holder complained, "When I hear these things about 'Let's make America great again,' I think to myself: 'Exactly when did you think America was great?'. . . . It takes us back to, I think, an American past that never really in fact existed with this notion of greatness."[2]

This viewpoint does not only suggest that America's history is replete with sins; every nation's history is. This viewpoint suggests that American greatness can *never* be achieved. It sets perfection as a goal, and then attributes failure to meet those goals to the very nature of

America and Americans. This is Disintegrationism at its finest. It sets a standard that can never be met, then blames the greatest country in history for failing to meet it.

Now, we live in a reactionary time; no doubt much of the hue and cry about "Make America Great Again" was due to pure reactionary feeling. But the belief that America was never great also has deeper philosophical roots. It is rooted in a rejection of American philosophy. It is rooted in a rejection of American culture. And it is rooted in a rejection of American history. The Disintegrationist view of America holds that reason does not lie at the root of human nature—and in fact, that there is no human nature at all absent a changeable and fixable society; that equality of rights before law is not enough, and that equality of result must be implemented; that individual rights must be put aside for the good of the community, with "positive rights" from the government replacing "negative rights" assured against government tyranny; that institutions designed to protect individual rights must be replaced with experts reigning from on high, and that the constitutional bargain ought to be dramatically rewritten. The Disintegrationist view of America holds that our culture of rights must be put aside in the name of safety and security: that religious institutions must be curbed, religious freedom restricted; that freedom of speech ought to be curtailed and a climate of speaking freely ought to be replaced by one of mass intimidation; that self-defense ought to be replaced by utter reliance on government; that economic freedom ought to be destroyed in the name of a more level society. The Disintegrationist view of America holds that American history was rooted in evil and never escaped it, and that all modern ills are simply manifestations of the dark beating heart implanted in the soil of the New World by the original European colonists and explorers.

In short, the Disintegrationist view of America holds that America must be uprooted: that the flag represents oppression rather than freedom, that America's fighting men and women are alternatively victims of a corrupt military-industrial complex or vicious monsters seeking blood and treasure abroad, that America's national anthem is the anthem of white supremacy and imperialism and exploitation.

THE SECOND AMERICAN REVOLUTION

Disintegrationists see themselves as new founders of the country. The country, in fact, is founded anew with each tweet, with each Facebook post, with each YouTube video decrying the evils of the United States. And the Disintegrationists need not pursue violence in the streets (although occasional violence is often tolerated and justified); instead, either they can convince Americans to rally to their virtue-signaling cause, or they can intimidate Americans into silence. They can convince Americans that true freedom lies in government privileges, withdrawn from some to be handed to others; that true duty lies in duty to the state; that the individual is troublesome, and that so long as most people feel free to pursue a modicum of pleasure, their true happiness is not endangered. There will be no battles of Lexington and Concord; there will be no shots fired, no guillotines in the streets. There will simply be a slow drift into enervation. Tocqueville foresaw this, too:

> I see an innumerable crowd of like and equal men who revolve on themselves without procuring the small and vulgar pleasures with which they fill their souls. Each of them, withdrawn and apart, is like a stranger to the destiny of all the others. . . . Above these an immense tutelary power is elevated, which alone takes charge of assuring their enjoyments and watching over their fate. It is absolute, detailed, regular, far-seeing, and mild. It would resemble paternal

power if, like that, it had for its object to prepare men for manhood; but on the contrary, it seeks only to keep them fixed irrevocably in childhood. . . . Thus, after taking each individual by turns in its powerful hands and kneading him as it likes, the sovereign extends its arms over society as a whole; it covers its surface with a network of small, complicated, painstaking, uniform rules through which the most original minds and the most vigorous souls cannot clear a way to surpass the crowd; it does not break wills, but it softens them, bends them, and directs them; it rarely forces one to act, but it constantly opposes itself to one's acting; it does not destroy, it prevents things from being born; it does not tyrannize, it hinders, compromises, enervates, extinguishes, dazes, and finally reduces each nation to being nothing more than a herd of timid and industrious animals of which the government is the shepherd.[3]

But it will not go this way.

That is because Americanism is rooted too deeply in Americans. We are too troublesome a people for this; we have spines, and will not be turned into jellyfish. More than that, we are not yet "like and equal" in the way Tocqueville foresaw. Only Disintegrationists believe that we are—or believe they can bully Americans into becoming so. Many Americans refuse to bow before Disintegrationist efforts to reduce and regularize them, their families, and their treasured institutions.

That refusal results in our supposedly "stupid" culture wars.

Our two visions of America—Unionist and Disintegrationist—are starkly different. They do not represent two ways of reaching the same goal. They represent, instead, two absolutely divergent goals. It is no wonder, then, that this deep-seated conflict breaks out into the open regularly. The tectonic plates of our self-identification are moving slowly and surely, snagging and releasing, grinding and tearing against each other—and with each tectonic clash, we feel the cultural earthquake. When an obscure actor named Jussie Smollett pretends to be victimized by a hate crime, and when we immediately file into political lines in reaction—that is a reflection of questions about America's history and America's present; when a marketing executive loses her job because she tweeted something stupid, and when we react instantaneously and passionately—that is a reflection of questions about our culture; when an actor declares a right to health care, and we all begin hashtagging—that is a reflection of questions about our philosophy.

These conflicts are exacerbated when the elements of American life that were heretofore seen as apolitical are politicized. It's a simple form of gaslighting to hijack formerly apolitical water-cooler arenas and institutions, then mock Unionist Americans as "triggered" when those apolitical areas are polarized. Those areas—areas like sports and Hollywood and music—were a respite from the deeper conflicts taking place across our nation. But as those islands of peace sink beneath the waves of controversy, Unionist Americans are *right* to be upset, right to feel assaulted. As the political and the cultural merge, Unionist Americans feel that the tectonic plates themselves will come crashing through the crust, revealing gaps far wider and more dangerous than the moderate-sized tremblers we feel every day.

Disintegrationists react by demanding ever more adherence, ever more control. They seek to dominate where they cannot persuade and convince. And so America will be left with two choices: submission or disintegration.

Unless.

Unless we remember.

Unless we learn to love; unless we learn to trust; unless we remember what unites us, rather than what divides us. We can start by remembering those who built our country—and why they built it.

THE GREAT CHOICE

Bringing up children is tough. It's tough because it's a job that constantly changes, day to day, moment to moment. It's a job of reaction, with the stimulus constantly morphing. How do you react when your newborn spits up on the brand-new carpet? How about when your toddler refuses to go to the potty? When your first grader demands TV before homework? And the questions only get tougher as the kids age. I'm not looking forward to dealing with my kids' questions about navigating the social wilds of high school, about going away to college, about dating, about marriage, about raising kids of their own.

Because parenting is an unending series of demands and choices, my tendency—like that of many parents, I'd think—is to become a control freak. It's difficult not to do so. After all, my kids want my attention full time (I've got three under the age of seven). The bathroom has become a bomb shelter—a respite for five minutes from the constant stream of "Daddy, can yous?" Rules and regulations are required to keep everyone moving and to keep everyone sane.

At the same time, it would be dishonest not to recognize that being a daddy is the most flattering work I can think of. It's flattering knowing that my kids care deeply what I think, that they want my approval, that they desire my guidance.

But as time goes on, my role is going to morph. My kids are going to get older, their choices more complex—and they aren't going to want my input as much. They're going to want to make their own choices. In reality, then, my job is to provide them the groundwork for making their own choices: to teach them the difference between right and wrong, the balance between justice and empathy, the requirement of personal responsibility.

And then there will come a time when they start making their own choices. And my job will be to let them be free. To let them go.

The hardest job in the world.

But also the second-most important. First comes teaching values; then comes letting my children apply those values in the real world. The greatest gift my parents gave me was the capacity to make my own choices. They trained me, and then they trusted me. I haven't always made the right choices; I won't always make the right choices. But that's what good parents do: equip their kids to the best of their abilities, make them responsible for the consequences of their own decisions, and then let them have an adventure. Our kids can only have their own adventures if we trust them—and if we trust ourselves to let go.

My children have been given the gift of the American adventure. We all have. But they can't have that adventure so long as their choices are circumscribed by the mob, by the state, by those who say that adventure is passé and that we all ought to seek permanent security from the collective. I want my kids to have rights against others, not merely privileges provided by others. I want my kids to grow up strong, independent, American.

I think most of us do.

I want my kids to have their adventure. That's the great joy of life: the adventure, choosing correctly in the face of adversity, overcoming obstacles and thriving. They can only have that adventure if I teach them to defend the individual rights our founders recognized—rights that preexist government, that adhere to them as individual human beings created in the image of God; if I teach them to engage in a culture of robust debate and to act with virtue in that culture; and if I remind them that they are inheritors of a grand and glorious American tradition, and that it is their responsibility to carry forward that tradition while widening its promises to more and more human beings.

They are Americans. And they have an obligation, as Americans, to understand American philosophy, American culture, and American history—and to become champions of them.

We all do.

And *that's* what should unify us.

CONCLUSION

Welcome to hell.

It's the middle of World War II. Millions are dying of brutal violence, genocide, and disease all over the globe. The date is April 1, 1945, and a young black man named Harry Stewart is flying. Stewart is a graduate of the Tuskegee Army Flying School, and a member of the famous Red Tails Fighter Group. He joined the military when he was eighteen years old.

Now, as flak bursts around him, Stewart looks down and sees the fighter planes of three of his fellow pilots on fire. One, smoking heavily, limps back toward Allied lines. A second crashes in a burst of flame in the fields below. The third plane crashes—but the pilot ejects.

Later, Stewart will find out what happened to that pilot, Walter Manning. A crowd of Austrians, agitated by SS troops, broke into the jail where Manning was being held. They beat him nearly to death. Then they hanged him from a lamppost.

As of this writing, Stewart is ninety-five years old, one of the last surviving Red Tails. He left a country of segregation. He returned to a country of segregation. But he fought for his country, and has no regrets. "The Axis Powers—I'm talking about Germany and Italy— they were no different. And when I say no different, [I mean] their idea of racial equality was even more egregious than what was in the United States," Stewart says. "My idea was to go ahead and fight for the United States of America. I was an invested citizen, one hundred percent American, and I felt as though that was my duty." But what made America worth fighting for? I asked him. Stewart said, "I guess the Constitution of the United States. You read that, and it's an absolutely beautiful document, but it wasn't being followed to the full extent. Since World War II, we have gotten closer to the ideal principles of that document."

Stewart says that while his fellow Americans may not have appreciated black Americans, all color disappeared in the sky. When the Red Tails pulled up alongside the B-17 and B-24 bombers, Stewart says, white bomber pilots breathed a sigh of relief. "We were like their guardian angels," he explains.

Stewart helped make hell into heaven.

In doing so, he stood with George Washington, Frederick Douglass, and Martin Luther King Jr.; he stood with Abraham Lincoln and Ronald Reagan; he stood with the American people.

On the day of the signing of the Declaration of Independence, John Hancock, the president of Congress, supposedly stated, "there must be no pulling different ways; we must all hang together." Benjamin Franklin replied, "yes, we must, indeed, all hang together, or most assuredly we shall all hang separately."

In 244 years, nothing has changed.

Yes, we are flawed.

Yes, we are cruel and generous and brutal and gentle and cynical and idealistic.

But most of all, we must not be enemies.

We *must* be Americans—Americans together—and we will be so again, when we are touched by the better angels of our nature.

ACKNOWLEDGMENTS

All books are the product of entire support systems: business networks, friends, family. This book has benefitted from those systems more than most.

Thanks to my editor at HarperCollins, Eric Nelson, a man unafraid to use the scalpel but kind enough to spare the ax. Thanks to my agent, Byrd Leavell, who has represented this work—and all my work—with passion and responsiveness far beyond the norm.

Thanks to Jeremy Boreing, who isn't merely the man behind the curtain at the *Daily Wire*, but also my best friend and an ideological and moral sounding board beyond compare. Thanks also to our common business partner, Caleb Robinson, a man of iron self-discipline and uncommon gentility.

Thanks to all our editors, writers, producers, marketing specialists, and the whole team at the *Daily Wire*, who make my job easier than it could be and far more fun than it should be.

Thanks to our broadcast partners at Westwood One, who were willing to think outside the box. It's rare to find creativity and risk taking in business relationships, but Westwood One has them both.

Thanks to our partners at Young America's Foundation, who make it a priority to bring conservatism to hundreds of thousands of students every year, and who give me a platform to spread the vision of small government and God-given rights to dozens of campuses every year.

Thanks to my longtime syndicators at Creators Syndicate, who took a chance on a seventeen-year-old kid. It's been two decades, and I'm still floored by that honor.

Some books are written over long periods of time, day by day; others are written at white heat. This book falls into the latter category. The problem, of course, with writing books at white heat is that you must dedicate inordinate hours at a stretch to writing them. To that end, thank you to my long-suffering wife, who picks up the slack and fills all the gaps. I married a wonder of a woman—she's a doctor, in case you didn't know—and she proves that every day with her optimism, dedication, and incredible ability to balance a husband, a medical practice, and three kids. When you get married, you take a leap into the unknown, trusting something great will happen. Best. Leap. Ever.

Thanks to my parents, who gave me the moral framework and ideological grounding to understand what makes America truly great— and who are there to watch the kids when my wife and I are both exhausted. Which is pretty much every day. For hours at a time.

And thanks to my kids, who are just the best: to my eldest daughter, who amazes me with her genius and brings me joy with her smile; to my son, an unending barrel of fun and a sensitive soul; and to my youngest daughter, who is just starting to enjoy the great adventure of growing up in this wild family. Good luck, kid!

Finally, thank you to God, who gave me the opportunity to live in the greatest country on earth and to defend the values that made it that way.

NOTES

INTRODUCTION

1. Kim Hart, "Exclusive Poll: Most Democrats See Republicans as Racist, Sexist," Axios.com, November 12, 2018, https://www.axios.com/poll-democrats-and-repub licans-hate-each-other-racist-ignorant-evil-99ae7afc-5a51-42be-8ee2-3959e43ce320 .html.

2. Carroll Doherty and Jocelyn Kiley, "Key Facts about Partisanship and Political Animosity in America," PewResearch.org, June 22, 2016, https://www.pewresearch .org/fact-tank/2016/06/22/key-facts-partisanship/.

3. John Wagner and Scott Clement, "'It's Just Messed Up': Most Think Political Divisions as Bad as Vietnam Era, New Poll Shows," *Washington Post*, October 28, 2017, https://www.washingtonpost.com/graphics/2017/national/democracy-poll/.

4. Daniel A. Cox, "Public Views of Political Compromise and Conflict and Partisan Misperceptions," AEI.org, October 2, 2019, https://www.aei.org/research-products /report/public-views-of-political-compromise-and-conflict-and-partisan-misper ceptions/.

5. "The Perception Gap," More In Common, 2019, https://perceptiongap.us/.

6. Lee Rainie, Scott Keeter, and Andrew Perring, "Trust and Distrust in America," Pew Research Center, July 22, 2019, https://www.people-press.org/2019/07/22 /the-state-of-personal-trust/.

7. Michelle Goldberg, "Bernie Could Win the Nomination. Should We Be Afraid?," *New York Times*, January 27, 2020, https://www.nytimes.com/2020/01/27/opinion/bernie-sanders-iowa-trump.html.

8. "9 DAYS TO IOWA: RALLY IN AMES WITH AOC," Bernie Sanders YouTube channel, January 25, 2020, https://www.youtube.com/watch?v=-joLGCz0Evw.

9. Max Roser, "Economic Growth," OurWorldInData.org, 2020, https://ourworldindata.org/economic-growth.

10. Zack Beauchamp, "600 Years of War and Peace, in One Amazing Chart," Vox.com, June 24, 2015, https://www.vox.com/2015/6/23/8832311/war-casualties-600-years.

11. Max Roser, Esteban Ortiz-Ospina, and Hannah Ritchie, "Life Expectancy," OurWorldInData.org, 2013, revised October 2019, https://ourworldindata.org/life-expectancy.

12. Max Fisher, "A Fascinating Map of the World's Most and Least Racially Tolerant Countries," *Washington Post*, May 15, 2013, https://www.washingtonpost.com/news/worldviews/wp/2013/05/15/a-fascinating-map-of-the-worlds-most-and-least-racially-tolerant-countries/.

13. Steven Pinker, "Has the Decline of Violence Reversed since *The Better Angels of Our Nature* Was Written?," StevenPinker.com, https://stevenpinker.com/files/pinker/files/has_the_decline_of_violence_reversed_since_the_better_angels_of_our_nature_was_written_2017.pdf.

14. Alex Thompson, "2020 Democrats Are Dramatically Changing the Way They Talk about Race," *Politico*, November 19, 2018, https://www.politico.com/story/2018/11/19/democrats-2020-race-identity-politics-strategy-1000249.

15. Deroy Murdock, "Beto's Gospel of Despair," *National Review*, September 13, 2019, https://www.nationalreview.com/2019/09/betos-gospel-of-despair/.

16. Steve Phillips, "The Next DNC Chair Must Abandon Color-Blind Politics," *The Nation*, January 19, 2017, https://www.thenation.com/article/the-next-dnc-chair-must-abandon-color-blind-politics/.

17. Kirsten Salyer, "The American Flag Was Banned at a Carolina High School," *Time*, August 30, 2016, https://time.com/4472433/american-flag-ban-south-carolina/.

18. Dartunorro Clark, "Democrat Beto O'Rourke, in Viral Video, Defends NFL Protests," NBCNews.com, August 23, 2018, https://www.nbcnews.com/politics/politics-news/democrat-beto-o-rourke-goes-viral-response-nfl-players-kneeling-n903176.

19. Eve Fairbanks, "The 'Reasonable Rebels,'" *Washington Post*, August 29, 2019, https://www.washingtonpost.com/outlook/2019/08/29/conservatives-say-weve-abandoned-reason-civility-old-south-said-that-too/?arc404=true.

20. Jeneen Interlandi, "Why Doesn't America Have Universal Healthcare? One Word: Race," *New York Times*, August 14, 2019, https://www.nytimes.com/interactive/2019/08/14/magazine/universal-health-care-racism.html.

21. Jamelle Bouie, "America Holds onto an Undemocratic Assumption from Its Founding: That Some People Deserve More Power than Others," *New York Times Maga-*

zine, August 14, 2019, https://www.nytimes.com/interactive/2019/08/14/magazine
/republicans-racism-african-americans.html.
22. Sandra Gonzalez, "Mario Lopez Apologizes for 'Ignorant' Comments About Parenting and Gender Identity," CNN.com, July 31, 2019, https://www.cnn.com/2019
/07/31/entertainment/mario-lopez-apology-trnd/index.html.

CHAPTER 1: THE AMERICAN PHILOSOPHY

1. Frederick Douglass, "What to the Slave Is the Fourth of July?," TeachingAmerican
History.org, July 5, 1852, https://teachingamericanhistory.org/library/document
/what-to-the-slave-is-the-fourth-of-july/.
2. Abraham Lincoln, Speech at Lewiston, Illinois, August 17, 1858, https://quod.lib
.umich.edu/l/lincoln/lincoln2/1:567?rgn=div1;view=fulltext.
3. Martin Luther King Jr., "'I Have A Dream' Address Delivered at the March on
Washington for Jobs and Freedom," King Institute at Stanford University, August 28,
1963, https://kinginstitute.stanford.edu/king-papers/documents/i-have-dream-address
-delivered-march-washington-jobs-and-freedom.
4. Thomas Jefferson, Letter to Henry Lee, May 8, 1825, https://teachingamerican
history.org/library/document/letter-to-henry-lee/.
5. See, for example, Virginia Schomp, *The Ancient Egyptians* (New York: Marshall
Cavendish, 2008), 41, and "Religion and Power: Divine Kingship in the Ancient World
And Beyond," Uchicago.edu, February 23–24, 2007, https://oi.uchicago.edu/research
/symposia/religion-and-power-divine-kingship-ancient-world-and-beyond-0.
6. Plato, *The Republic* (New York: Basic Books, 1968), 353c–353e.
7. Aristotle, *Nichomachean Ethics* (Chicago: University of Chicago Press, 2011),
1098a.
8. Richard Tarnas, *The Passion of the Western Mind* (New York: Ballantine Books,
1991), 175.
9. Thomas P. Rausch, *Reconciling Faith and Reason: Apologists, Evangelists, and
Theologians in a Divided Church* (Collegeville, MN: Liturgical Press, 2000), 12.
10. Richard H. Cox, "Hugo Grotius," in Leo Strauss and Joseph Crowley, eds., *History of Political Philosophy*, 3rd ed. (Chicago: University of Chicago Press, 1987), 389.
11. *De Iure Praedae Commentarius*, trans. Gwladys L. Williams and Walter H. Zeydel
(Carnegie Endowment for International Peace, Oxford University Press, 1950), 1:18,
as quoted in "Hugo Grotius, The Rights of War and Peace (1625)," Online Library of
Liberty, https://oll.libertyfund.org/pages/grotius-war-peace.
12. John Locke, *Second Treatise of Government* (1690), Section 22, https://www
.gutenberg.org/files/7370/7370-h/7370-h.htm.
13. Locke, *Second Treatise of Government*, Section 6, https://www.gutenberg.org
/files/7370/7370-h/7370-h.htm.
14. Adam Smith, *An Inquiry into the Nature and Causes of the Wealth of Nations* (London: Methuen, 1776), book IV, chapter 9, sec. 50.
15. C. Bradley Thompson, *America's Revolutionary Mind* (New York: Encounter
Books, 2019), 38.

16. Ibid., 43, 55.

17. Ibid., 181

18. "Jefferson's 'original Rough draft' of the Declaration of Independence," *The Papers of Thomas Jefferson*, vol. 1, *1760–1776* (Princeton, NJ: Princeton University Press, 1950), 423–28, https://jeffersonpapers.princeton.edu/selected-documents/jefferson%E2%80%99s-%E2%80%9Coriginal-rough-draught%E2%80%9D-declaration-independence.

19. Aristotle, *Politics* (Chicago: University of Chicago Press, 2013), book 1, chapter 5, 1254b.

20. "Magna Carta: Muse and Mentor," Library of Congress, November 6, 2014–January 19, 2015, https://www.loc.gov/exhibits/magna-carta-muse-and-mentor/magna-carta-and-the-us-constitution.html.

21. Locke, *Second Treatise of Government*, Section 6, https://www.gutenberg.org/files/7370/7370-h/7370-h.htm.

22. Thomas Paine, *Common Sense*, February 14, 1776, https://www.learner.org/workshops/primarysources/revolution/docs/Common_Sense.pdf.

23. Thompson, *America's Revolutionary Mind*, 118.

24. Pauline Maier, "The Strange History of 'All Men Are Created Equal,'" *Washington and Lee Law Review* 56, no. 3 (June 1, 1999), https://scholarlycommons.law.wlu.edu/cgi/viewcontent.cgi?article=1547&context=wlulr.

25. Massachusetts Constitution, 1780, https://malegislature.gov/laws/constitution.

26. Thompson, *America's Revolutionary Mind*, 143.

27. Abraham Lincoln, *Speeches and Writings 1832–1858* (New York: Library of America, 1989), 398–400.

28. Booker T. Washington, *The Booker T. Washington Papers*, vol. 12, *1912–1914* (Urbana: University of Illinois Press, 1982), 260.

29. Deuteronomy 17:14–20.

30. 1 Samuel 8:10–18.

31. Plato, *The Republic*, 473c-d.

32. Aristotle, *Politics*, book II, chapter V, http://classics.mit.edu/Aristotle/politics.2.two.html.

33. Cicero, *De Re Publica* (Cambridge, MA: Loeb Classical Library, 2000), 203.

34. Pascal Daudin, "The Thirty Years' War: The First Modern War?," *Humanitarian Law and Policy*, May 23, 2017, https://blogs.icrc.org/law-and-policy/2017/05/23/thirty-years-war-first-modern-war/.

35. Martti Koskenniemi, "Imagining the Rule of Law: Rereading the Grotian 'Tradition,'" *European Journal of International Law* 30, no. 1 (2019), https://academic.oup.com/ejil/article/30/1/17/5498077#136043457.

36. John Locke, *Two Treatises of Government* (London: Awnsham Churchill, 1689), book 2, sec. 57.

37. Ibid., book 2, sec. 222.

38. George A. Peek Jr., ed., *The Writings of John Adams* (Indianapolis: Hackett, 2003), 154.

39. James Madison, *Federalist No. 10* (1787).

40. Charlie Spiering, "Obama: The Government Is Us and We're Doing Things Right," WashingtonExaminer.com, July 8, 2013, https://www.washingtonexaminer.com/video-obama-the-government-is-us-and-were-doing-things-right.

41. Mark Puls, *Samuel Adams: Father of the American Revolution* (New York: Palgrave Macmillan, 2006), 36.

42. Henry Stephens Randall, *The Life of Thomas Jefferson* (New York: Derby & Jackson, 1858), 117.

43. Douglass, "What to the Slave Is the Fourth of July?"

44. James Madison, *Federalist No. 51* (1788).

45. James Madison, *Federalist No. 45* (1788).

46. James Madison, *Federalist No. 14* (1787).

47. Alexander Hamilton, *Federalist No. 84* (1788).

48. James Madison, *Federalist No. 51* (1788).

49. Alexander Hamilton, *Federalist No. 70* (1788).

50. Alexander Hamilton, *Federalist No. 69* (1788).

51. Alexander Hamilton, *Federalist No. 78* (1788).

52. Ibid.

53. Carson Holloway, "Against Judicial Supremacy: The Founders and the Limits on the Courts," Heritage.org, January 25, 2019, https://www.heritage.org/courts/report/against-judicial-supremacy-the-founders-and-the-limits-the-courts.

54. Randy E. Barnett, *Restoring the Lost Constitution* (Princeton, NJ: Princeton University Press, 2004), 61–62.

55. James Madison, *Federalist No. 10* (1787).

56. James Madison, *Federalist No. 51* (1788).

57. Abraham Lincoln, "Fragment on the Constitution and the Union," January 1861, *Collected Works of Abraham Lincoln*, vol. 4, https://quod.lib.umich.edu/l/lincoln/lincoln4/1:264?rgn=div1;view=fulltext.

CHAPTER 2: DISINTEGRATING AMERICAN PHILOSOPHY

1. "Text of President Obama's Speech in Hiroshima, Japan," NYTimes.com, May 27, 2016, https://www.nytimes.com/2016/05/28/world/asia/text-of-president-obamas-speech-in-hiroshima-japan.html.

2. Jonah Goldberg, "Looking to Obama for Redemption," NYPost.com, June 7, 2008, https://nypost.com/2008/06/07/looking-to-obama-for-redemption/.

3. Cathleen Falsani, "Transcript: Barack Obama and the God Factor Interview," Sojo.net, February 21, 2012, https://sojo.net/articles/transcript-barack-obama-and-god-factor-interview.

4. "Inaugural Address by President Barack Obama," The White House, January 21, 2013, https://obamawhitehouse.archives.gov/the-press-office/2013/01/21/inaugural-address-president-barack-obama.

5. Barack Obama, Twitter, October 24, 2016, https://twitter.com/barackobama/status/790573868335308801?lang=en.

6. "Inaugural Address by President Barack Obama."

7. Lyndon B. Johnson, "Commencement Address at Howard University: 'To Fulfill These Rights,'" June 4, 1965, https://teachingamericanhistory.org/library/document /commencement-address-at-howard-university-to-fulfill-these-rights/.

8. Spiering, "Obama: The Government Is Us and We're Doing Things Right."

9. Joel Gehrke, "DNC: 'Government Is the Only Thing That We All Belong To,'" WashingtonExaminer.com, September 4, 2012, https://www.washingtonexaminer .com/dnc-government-is-the-only-thing-that-we-all-belong-to.

10. James Madison, *Federalist No. 51* (1788).

11. "Remarks by the President at The Ohio State University," White House, May 5, 2013, https://obamawhitehouse.archives.gov/the-press-office/2013/05/05/remarks -president-ohio-state-university-commencement.

12. Roger D. Masters, *The Political Philosophy of Rousseau* (Princeton, NJ: Princeton University Press, 1976), 5.

13. Jean Jacques Rousseau, *On the Origin of the Inequality of Mankind* (1754), https:// www.marxists.org/reference/subject/economics/rousseau/inequality/ch02.htm.

14. Karl Marx, "Private Property and Communism," 1844, https://www.marxists .org/archive/marx/works/1844/manuscripts/comm.htm.

15. Erich Fromm, "Marx's Concept of Man," 1961, https://www.marxists.org/archive /fromm/works/1961/man/ch04.htm.

16. Tiffany Jones Miller, "Transforming Formal Freedom into Effective Freedom: Dewey, the New Deal, and the Great Society," in Ronald Pestritto and Thomas West, eds., *Modern America and the Legacy of the Founding* (Lanham, MD: Lexington Books, 2007), 178.

17. Woodrow Wilson, *Selected Addresses and Public Papers of Woodrow Wilson* (New York: Boni & Liveright, 1918), 48.

18. John Dewey, *John Dewey: The Later Works, 1925–1953* (Carbondale: Southern Illinois University Press, 1988), 291–93.

19. Ian Schwartz, "Sanders Campaign Organizer: Free Education, Gulags Needed to 'Re-Educate' You to Not Be a 'F*cking Nazi,'" RealClearPolitics.com, January 14, 2020, https://www.realclearpolitics.com/video/2020/01/14/sanders_campaign_organizer _free_education_gulags_needed_to_re-educate_you_not_to_be_a_fcking_nazi .html.

20. Steven Pinker, "From 'The Blank Slate,'" *Discover* (October 2002), 34–40.

21. "How Do We Bring Equality to Data Ownership and Usage?," Wired.com, November 8, 2019, https://www.wired.com/story/laura-boykin-malkia-devich-cyril -data-privacy-wired25/.

22. Donna M. Hughes, "Significant Differences: The Construction of Knowledge, Objectivity, and Dominance," *Women's Studies International Forum* 18, no. 4 (July–August 1995): 395–406.

23. Craig S. Smith, "Dealing with Bias in Artificial Intelligence," NYTimes.com, November 19, 2019, https://www.nytimes.com/2019/11/19/technology/artificial -intelligence-bias.html.

24. Allum Bokhari, "Lawsuit: Google Instructed Managers That 'Individual Achieve-

ment' and 'Objectivity' Were Examples of 'White Dominant Culture,'" Breitbart .com, April 18, 2018, http://www.breitbart.com/tech/2018/04/18/lawsuit-google -instructed-managers-that-individual-achievement-and-objectivity-were-examples -of-white-dominant-culture/.

25. John Sexton, "Professor Notes Men Are Taller Than Women on Average, SJWs Storm Out Angrily," HotAir.com, March 14, 2018, https://hotair.com/archives /john-s-2/2018/03/14/professor-points-men-taller-women-average-sjws-storm -angrily/.

26. Heather Heying, "Grievance Studies vs. the Scientific Method," Medium.com, November 1, 2018, https://medium.com/@heyingh/grievance-studies-goes-after -the-scientific-method-63b6cfd9c913.

27. Jillian Kay Melchior, "Fake News Comes to Academia," WSJ.com, October 5, 2018, https://www.wsj.com/articles/fake-news-comes-to-academia-1538520950.

28. Colleen Flaherty, "Blowback Against a Hoax," InsideHigherEd.com, January 8, 2019, https://www.insidehighered.com/news/2019/01/08/author-recent-academic -hoax-faces-disciplinary-action-portland-state.

29. Kathleen Doheny, "Boy or Girl? Fetal DNA Tests Often Spot On," WebMD.com, August 9, 2011, https://www.webmd.com/baby/news/20110809/will-it-be-a-boy -or-girl-fetal-dna-tests-often-spot-on#1.

30. Marilynn Marchione, "Nurse Mistakes Pregnant Transgender Man as Obese. Then, the Man Births a Stillborn Baby," Associated Press, May 16, 2019, https://www .usatoday.com/story/news/health/2019/05/16/pregnant-transgender-man-births -stillborn-baby-hospital-missed-labor-signs/3692201002/.

31. "AMA Adopts New Policies During First Day of Voting at Interim Meeting," American Medical Association, November 19, 2019, https://www.ama-assn.org/press -center/press-releases/ama-adopts-new-policies-during-first-day-voting-interim -meeting.

32. Amanda Prestigiacomo, "Jury Rules Against Texas Dad Fighting 7-Year-Old Son's Gender Transition," DailyWire.com, October 22, 2019, https://www.dailywire.com /news/jury-rules-against-texas-dad-trying-to-save-7-year-old-son-from-gender -transition-potential-castration.

33. Ben Shapiro, "A Brown University Researcher Released a Study about Teens Imitating Their Peers by Turning Trans. The Left Went Insane. So Brown Caved," DailyWire.com, August 28, 2018, https://www.dailywire.com/news/brown-uni versity-researcher-released-study-about-ben-shapiro.

34. Declaration of the Rights of Man, 1789, http://avalon.law.yale.edu/18th_century /rightsof.asp.

35. Karl Marx, *Critique of the Gotha Program* (N.p.: Wildside Press, 2008), 26–27.

36. Dewey, *John Dewey: The Later Works, 1925–1953*, 21–22.

37. Woodrow Wilson, "Address to the Jefferson Club of Los Angeles (1911)," in Scott J. Hammond et al., eds., *Classics of American Political Thought and Constitutional Thought*, vol. 2 (Indianapolis/Cambridge: Hackett, 2007), 323–24.

38. Woodrow Wilson, "The Author and Signers of the Declaration of Independence,"

http://cdn.constitutionreader.com/files/pdf/coursereadings/Con201_Readings
_Week2_AuthorandSigners.pdf.

39. Franklin D. Roosevelt, "State of the Union Address (1944)," https://teaching
americanhistory.org/library/document/state-of-the-union-address-3/.

40. Saranac Hale Spencer, "Ruth Bader Ginsburg Taken Way Out of Context," Fact
Check.org, December 13, 2018, https://www.factcheck.org/2018/12/ruth-bader
-ginsburg-taken-way-out-of-context/.

41. Tara Golshan, "Read: Bernie Sanders Defines His Vision for Democratic Socialism in the United States," Vox.com, June 12, 2019, https://www.vox.com/2019
/6/12/18663217/bernie-sanders-democratic-socialism-speech-transcript.

42. Ben Shapiro, "Obama Returns, Gives a Speech Reminding Americans of Why
Trump Is President," DailyWire.com, September 7, 2018 https://www.dailywire
.com/news/obama-returns-gives-speech-reminding-americans-why-ben-shapiro.

43. John C. Calhoun, "Speech on the Oregon Bill," June 27, 1848, https://teaching
americanhistory.org/library/document/oregon-bill-speech/.

44. George F. Will, "The Liberals Who Loved Eugenics," WashingtonPost.com,
March 8, 2017 https://www.washingtonpost.com/opinions/the-liberals-who-loved
-eugenics/2017/03/08/0cc5e9a0-0362-11e7-b9fa-ed727b644a0b_story.html.

45. Jonah Goldberg, *Suicide of the West* (New York: Crown Forum, 2018).

46. John Rawls, *A Theory of Justice* (Cambridge, MA: Harvard University Press,
1971), 15.

47. John Rawls, *Justice as Fairness: A Restatement* (Cambridge, MA: Harvard University Press, 2001), 42.

48. Paul Krugman, "More Thoughts on Equality of Opportunity," NYTimes.com,
January 11, 2011, https://krugman.blogs.nytimes.com/2011/01/11/more-thoughts
-on-equality-of-opportunity/.

49. Theodore R. Johnson, "How Conservatives Turned the 'Color-Blind Constitution' Against Racial Progress," TheAtlantic.com, November 19, 2019, https://www
.theatlantic.com/ideas/archive/2019/11/colorblind-constitution/602221/.

50. Kimberlé Crenshaw, "Why Intersectionality Can't Wait," *Washington Post*, September 24, 2015, https://www.washingtonpost.com/news/in-theory/wp/2015/09/24
/why-intersectionality-cant-wait/?noredirect=on&utm_term=.179ecf062277.

51. Erika Smith, "A Crying Brett Kavanaugh. This Is What White Male Privilege
Looks Like," McClatchy DC, September 27, 2018 https://www.mcclatchydc.com
/opinion/article219146885.html; J. Mills Thorton III, *Dividing Lines: Municipal Politics and the Struggle for Civil Rights in Montgomery, Birmingham, and Selma* (Tusca-
loosa: University of Alabama Press, 2002), 68.

52. Thorton, *Dividing Lines*, 68.

53. Lawrence H. Summers, "Remarks at NBER Conference on Diversifying the Science & Engineering Workforce," Office of the President of Harvard University, January 14, 2005, https://web.archive.org/web/20080130023006/http://www.president
.harvard.edu/speeches/2005/nber.html.

54. Michael Harriot, "Pete Buttigieg Is a Lying MF," TheRoot.com, November 25, 2019, https://www.theroot.com/pete-buttigieg-is-a-lying-mf-1840038708.

55. Heather MacDonald, "How Identity Politics Is Harming the Sciences," City-Journal .org, Spring 2018, https://www.city-journal.org/html/how-identity-politics-harm ing-sciences-15826.html.

56. Eliza Shapiro, "Beacon High School Is Half White. That's Why Students Walked Out," NYTimes.com, December 2, 2019, https://www.nytimes.com/2019/12/02 /nyregion/nyc-beacon-high-school-walkout.html?action=click&module=Top%20 Stories&pgtype=Homepage.

57. Nell Greenfieldboyce, "Academic Science Rethinks All-Too-White 'Dude Walls' of Honor," NPR.org, August 25, 2019, https://www.npr.org/sections/health-shots /2019/08/25/749886989/academic-science-rethinks-all-too-white-dude-walls-of -honor.

58. Mary Pickering, *Auguste Comte: An Intellectual Biography*, vol. 1 (Cambridge: Cambridge University Press, 1993), 211.

59. John R. Shook and James A. Good, *John Dewey's Philosophy of Spirit, with the 1897 Lecture on Hegel* (New York: Fordham University Press, 2010), 29.

60. Matthew Festenstein, *Pragmatism and Political Theory: From Dewey to Rorty* (Chicago: University of Chicago Press, 1997), 65.

61. K. Sabeel Rahman, *Democracy Against Domination* (Oxford: Oxford University Press, 2017), 175.

62. Woodrow Wilson, "Socialism and Democracy," August 22, 1887, https://teaching americanhistory.org/library/document/socialism-and-democracy/.

63. As quoted in Ronald J. Pestritto, "Woodrow Wilson and the Rejection of the Founders' Principles," Hillsdale.edu, https://online.hillsdale.edu/document.doc?id =313.

64. Woodrow Wilson, "The Study of Administration," *Political Science Quarterly* 2, no. 2 (June 1887).

65. Barack Obama, "Remarks by the President Presenting New Management Agenda," The White House, July 8, 2013, https://obamawhitehouse.archives.gov/the-press -office/2013/07/08/remarks-president-presenting-new-management-agenda.

66. Myron Magnet, *The Founders at Home* (New York: Norton, 2014), 358.

67. *Wickard v. Filburn*, 317 US 111 (1942).

68. Eric Segall and Aaron E. Carroll, "Health Care and Constitutional Chaos," *Stanford Law Review*, May 2012, https://www.stanfordlawreview.org/online/health -care-and-constitutional-chaos/.

69. Tom McCarthy, "Senate Approves Change to Filibuster Rule after Repeated Republican Blocks," *Guardian* (UK), November 21, 2013, https://www.theguardian.com /world/2013/nov/21/harry-reid-senate-rules-republican-filibusters-nominations.

70. David Faris, "Elizabeth Warren Is Right: Abolish the Filibuster," TheWeek.com, September 13, 2019, https://theweek.com/articles/864525/elizabeth-warren-right -abolish-filibuster.

71. Jamelle Bouie, "The Senate Is as Much of a Problem as Trump," *New York Times*, May 10, 2019, https://www.nytimes.com/2019/05/10/opinion/sunday/senate-dem ocrats-trump.html.

72. Woodrow Wilson, "The President of the United States," 1908, https://online .hillsdale.edu/file/presidency/lecture-4/The-President-Of-the-United-States -Woodrow-Wilson-Pgs.-649-660.pdf.

73. "Total Pages Published in the Code of Federal Regulations (1950–2018)," George Washington University Regulatory Studies Center, https://regulatorystudies .columbian.gwu.edu/sites/g/files/zaxdzs1866/f/image/GW%20Reg%20Studies %20-%20Pages%20Published%20in%20the%20CFR%20-%206.12.19.png.

74. "Executive Branch Civilian Employment Since 1940," Office of Personnel and Management, https://www.opm.gov/policy-data-oversight/data-analysis-documen tation/federal-employment-reports/historical-tables/executive-branch-civilian -employment-since-1940/.

75. Stephen Moore, "The Growth of Government in America," FEE.org, April 1, 1993, https://fee.org/articles/the-growth-of-government-in-america/.

76. "Federal Net Outlays as Percent of Gross Domestic Product," Federal Reserve Bank of St. Louis, https://fred.stlouisfed.org/series/FYONGDA188S.

77. Alex Kozinski, "My Pizza with Nino," *Cardozo Law Review* 12 (1991), 1583, 1588–89.

78. John Hasnas, "The 'Unseen' Deserve Empathy, Too," *Wall Street Journal*, May 29, 2009, https://www.cato.org/publications/commentary/unseen-deserve-empathy-too.

79. Evan D. Bernick, "'Uncommonly Silly'—and Correctly Decided: The Right and Wrong of *Griswold v. Connecticut* and Why It Matters Today," FedSoc.org, April 18, 2017, https://fedsoc.org/commentary/blog-posts/uncommonly-silly-and-correctly -decided-the-right-and-wrong-of-griswold-v-connecticut-and-why-it-matters-today.

80. *Planned Parenthood v. Casey*, 505 U.S. 833 (1992).

81. *Obergefell v. Hodges*, 576 US __ (2015).

82. Ilya Somin, "How Liberals Learned to Love Federalism," WashingtonPost.com, July 12, 2019, https://www.washingtonpost.com/outlook/how-liberals-learned -to-love-federalism/2019/07/12/babd9f52-8c5f-11e9-b162-8f6f41ec3c04_story .html.

CHAPTER 3: THE AMERICAN CULTURE

1. Edmund Burke, *Reflections on the Revolution in France* (New York: Oxford University Press, 1999).

2. Alexander Hamilton, *Federalist No. 84*, https://avalon.law.yale.edu/18th_century /fed84.asp.

3. Robert Shibley, "For the Fourth: Ben Franklin on Freedom of Speech—50 Years Before the Constitution," July 4, 2016, https://www.thefire.org/for-the-fourth-ben -franklin-on-freedom-of-speech-50-years-before-the-constitution/.

4. George Washington, Address to the Officers of the Army, March 15, 1783, https://www.mountvernon.org/library/digitalhistory/quotes/article/for-if-men -are-to-be-precluded-from-offering-their-sentiments-on-a-matter-which-may

-involve-the-most-serious-and-alarming-consequences-that-can-invite-the-consid
eration-of-mankind-reason-is-of-no-usc-to-us-the-freedom-of-speech-may-be-taken
-away-and-dumb-/.

5. John Adams, *The Political Writings of John Adams* (Washington, DC: Regnery, 2000), 13.

6. *Abrams v. United States* (250 U.S. 630).

7. John Adams, Letter to Massachusetts Militia, October 11, 1798, https://founders .archives.gov/documents/Adams/99-02-02-3102.

8. George Washington, "First Inaugural Address," April 30, 1789, https://www .archives.gov/exhibits/american_originals/inaugtxt.html.

9. James Madison, "Memorial and Remonstrance against Religious Assessment— Full Text" (1785), https://billofrightsinstitute.org/founding-documents/primary -source-documents/memorial-and-remonstrance/.

10. Thomas Jefferson, Letter to Thomas Leiper, January 21, 1809, https://founders .archives.gov/documents/Jefferson/99-01-02-9606.

11. Alexis de Tocqueville, "On the Use That the Americans Make of Association in Civil Life," trans. Harvey Mansfield and Delba Winthrop, from *Democracy in America*, http://www.press.uchicago.edu/Misc/Chicago/805328.html.

12. Alexis de Tocqueville, *Democracy in America*, trans. Harvey Mansfield and Delba Winthrop (Chicago: University of Chicago Press, 2000).

13. James Madison, "Memorial and Remonstrance against Religious Assessment— Full Text" (1785).

14. "Thomas Jefferson and the Virginia Statute for Religious Freedom," Virginia Museum of History & Culture, https://www.virginiahistory.org/collections-and -resources/virginia-history-explorer/thomas-jefferson.

15. Jonathan Evans, "US Adults Are More Religious than Western Europeans," Pew Research.org, September 5, 2018, https://www.pewresearch.org/fact-tank/2018 /09/05/u-s-adults-are-more-religious-than-western-europeans/.

16. David Masci and Claire Gecewicz, "Share of Married Adults Varies Widely Across US Religious Groups," PewResearch.org, March 19, 2018, https://www .pewresearch.org/fact-tank/2018/03/19/share-of-married-adults-varies-widely -across-u-s-religious-groups/.

17. Michael Lipka, "Mormons More Likely to Marry, Have More Children than Other US Religious Groups," PewResearch.org, May 22, 2015, https://www.pewresearch .org/fact-tank/2015/05/22/mormons-more-likely-to-marry-have-more-children -than-other-u-s-religious-groups/.

18. Alexander Hamilton, *Federalist No. 29* (1788), https://avalon.law.yale.edu/18th _century/fed29.asp.

19. "No freeman shall be debarred the use of arms," Monticello.org, https://www .monticello.org/site/research-and-collections/no-freeman-shall-be-debarred-use -arms.

20. David Harsanyi, *First Freedom: A Ride Through America's Enduring History with the Gun, from the Revolution to Today* (New York: Simon & Schuster, 2018), 63.

21. James Lindgren and Justin Heather, "Counting Guns in Early America," *William & Mary Law Review* 43, no. 5 (2002), http://scholarship.law.wm.edu/cgi/viewcontent .cgi?article=1489&context=wmlr.

22. Harsanyi, *First Freedom*, 68.

23. Noah Shusterman, "What the Second Amendment Really Meant to the Founders," *Washington Post*, February 12, 2008, https://www.washingtonpost.com/news /made-by-history/wp/2018/02/22/what-the-second-amendment-really-meant-to -the-founders/.

24. Zach Weismueller, "Gun Rights, Civil Rights," Reason.com, June 2015, https:// reason.com/2015/04/30/gun-rights-civil-rights/.

25. Charles E. Cobb, *This Nonviolent Stuff'll Get You Killed* (New York: Basic Books, 2014).

26. Locke, *Second Treatise of Government*.

27. James Madison, *Federalist No. 10* (1787).

28. Friedrich Hayek, *The Fatal Conceit* (London: Routledge, 1988), 7.

29. "North Korea's Economy Grew at 3.7% in 2017, Pyongyang Professor Estimates," *Japan Times*, October 13, 2018, https://www.japantimes.co.jp/news/2018/10/13/asia -pacific/north-koreas-economy-grew-3-7-2017-pyongyang-professor-estimates/#.XR Ebd9NKjOQ.

30. "GDP Per Capita," WorldBank.org, https://data.worldbank.org/indicator/NY .GDP.PCAP.CD?locations=JP-KR.

31. "Nine Charts Which Tell You All You Need to Know About North Korea," BBC .com, September 26, 2017, https://www.bbc.com/news/world-asia-41228181.

32. "Danish PM in US: Denmark Is Not Socialist," *The Local* (Denmark), November 1, 2015, https://www.thelocal.dk/20151101/danish-pm-in-us-denmark-is-not -socialist.

33. Rich Lowry, "Sorry, Bernie—Scandinavia Is No Socialist Paradise After All," NYPost.com, October 19, 2015, http://nypost.com/2015/10/19/sorry-bernie-scan dinavia-is-no-socialist-paradise-after-all/.

34. Tocqueville, *Democracy in America*.

35. H. G. Wells, *The Future in America: A Search After Realities* (N.p.: Musaicum Books, 2017).

36. Tocqueville, *Democracy in America*.

CHAPTER 4: THE DISINTEGRATIONIST CULTURE

1. Herbert Marcuse, "Repressive Tolerance" (1965), https://www.marcuse.org /herbert/pubs/60spubs/65repressivetolerance.htm.

2. Eve Fairbanks, "The 'Reasonable Rebels,'" *Washington Post*, August 29, 2019, https://www.washingtonpost.com/outlook/2019/08/29/conservatives-say-weve -abandoned-reason-civility-old-south-said-that-too/?arc404=true.

3. Kevin D. Williamson, *The Smallest Minority* (Washington, DC: Gateway Editions: 2019)

4. "ACLU Case Selection Guidelines: Conflicts between Competing Values or Prior-

ities," WSJ.com, June 21, 2018, http://online.wsj.com/public/resources/documents/20180621ACLU.pdf?mod=article_inline.

5. Ben Shapiro, "INSANE: ACLU Now Opposes Accused Having So Many Rights, All to Bash Trump," DailyWire.com, November 16, 2018, https://www.dailywire.com/news/insane-aclu-now-opposes-accused-students-having-so-ben-shapiro.

6. Alexis de Tocqueville, *Democracy in America* (London: Longmans, Green, 1875), 268–69.

7. Ibid.

8. Jon Ronson, "How One Stupid Tweet Blew Up Justine Sacco's Life," NYTimes.com, February 15, 2015, https://www.nytimes.com/2015/02/15/magazine/how-one-stupid-tweet-ruined-justine-saccos-life.html.

9. Associated Press, "Mozilla CEO Resignation Raises Free-Speech Issues," USA Today.com, April 4, 2014, https://www.usatoday.com/story/news/nation/2014/04/04/mozilla-ceo-resignation-free-speech/7328759/.

10. Caitlin Flanagan, "The Media Botched the Covington Catholic Story," The Atlantic.com, January 23, 2019, https://www.theatlantic.com/ideas/archive/2019/01/media-must-learn-covington-catholic-story/581035/.

11. Adam Johnson, "Sarah Silverman: Anger Over Political Correctness Is a Sign of Being Old," Alternet.org, September 16, 2015, https://www.alternet.org/2015/09/sarah-silverman-anger-over-political-correctness-sign-being-old-video/.

12. Francesca Bacardi, "Sarah Silverman Fired from New Movie for Blackface Photo," PageSix.com, August 12, 2019, https://pagesix.com/2019/08/12/sarah-silverman-fired-from-new-movie-for-blackface-photo/.

13. Elahe Izadi, "Hannah Gadsby Broke Comedy. So What's She Building Now?," WashingtonPost.com, July 12, 2019, https://www.washingtonpost.com/lifestyle/style/hannah-gadsby-broke-comedy-so-whats-she-building-now/2019/07/11/3f720124-a27d-11e9-bd56-eac6bb02d01d_story.html.

14. Aude White, "On the Cover: In Conversation with Jimmy Kimmel," *New York*, October 30–November 12, 2017, http://nymag.com/press/2017/10/on-the-cover-in-conversation-with-jimmy-kimmel.html.

15. Ben Shapiro, "Media Matters Makes America a Worse Place, One Bad Faith Hit at a Time," DailyWire.com, March 11, 2019, https://www.dailywire.com/news/media-matters-makes-america-worse-place-one-bad-ben-shapiro.

16. Jennifer Medina, "Bernie Sanders Retracts Endorsement of Cenk Uygur After Criticism," NYTimes.com, December 13, 2019, https://www.nytimes.com/2019/12/13/us/politics/bernie-sanders-cenk-uygur.html.

17. Michael S. Roth, "Don't Dismiss 'Safe Spaces,'" NYTimes.com, August 29, 2019, https://www.nytimes.com/2019/08/29/opinion/safe-spaces-campus.html.

18. "Spotlight on Speech Codes 2020," TheFire.org, https://www.thefire.org/resources/spotlight/reports/spotlight-on-speech-codes-2020/.

19. Greg Lukianoff and Jonathan Haidt, "The Coddling of the American Mind," TheAtlantic.com, September 2015, https://www.theatlantic.com/magazine/archive/2015/09/the-coddling-of-the-american-mind/399356/.

20. Celine Ryan, "POLL: Most Young Americans Support 'Hate Speech' Exemption in First Amendment," CampusReform.org, October 25, 2019, https://www.campus reform.org/?ID=13916.

21. Julia Alexander, "YouTube Revokes Ads from Steven Crowder Until He Stops Linking to His Homophobic T-shirts," TheVerge.com, June 5, 2019, https://www.theverge.com/2019/6/5/18654196/steven-crowder-demonetized-carlos-maza-youtube-homophobic-language-ads.

22. "YouTube Bans 'Malicious Insults and Veiled Threats,'" BBC.com, December 11, 2019, https://www.bbc.com/news/technology-50733180.

23. Matt Drake, "Researcher Who Lost Job for Tweeting 'Men Cannot Turn into Women' Loses Employment Tribunal," Independent.co.uk, December 19, 2019, https://www.independent.co.uk/news/uk/home-news/maya-forstater-transgender-test-case-equalities-act-employment-tribunal-a9253211.html.

24. Aja Romano, "J. K. Rowling's Latest Tweet Seems Like Transphobic BS. Her Fans Are Heartbroken," Vox.com, December 19, 2019, https://www.vox.com/culture/2019/12/19/21029852/jk-rowling-terf-transphobia-history-timeline.

25. Eugene Volokh, "You Can Be Fined for Not Calling People 'Ze' or 'Hir,' If That's the Pronoun They Demand That You Use," WashingtonPost.com, May 17, 2016, https://www.washingtonpost.com/news/volokh-conspiracy/wp/2016/05/17/you-can-be-fined-for-not-calling-people-ze-or-hir-if-thats-the-pronoun-they-demand-that-you-use/.

26. Richard Stengel, "Why America Needs a Hate Speech Law," WashingtonPost.com, October 29, 2019, https://www.washingtonpost.com/opinions/2019/10/29/why-america-needs-hate-speech-law/.

27. Karl Marx, *Critique of Hegel's "Philosophy of Right"* (Cambridge: Cambridge University Press, 1970), 131.

28. Karl Marx and Frederick Engels, *The Communist Manifesto* (Chicago: Haymarket Books, 2005), 66–67.

29. Danielle Kurtzleben, "We're All Getting Married Too Late," USNews.com, February 14, 2014, https://www.usnews.com/news/blogs/data-mine/2014/02/14/think-were-all-getting-married-super-late-think-again.

30. Ben Wattenberg, *The First Measured Century*, "Family," PBS.org, https://www.pbs.org/fmc/book/4family9.htm.

31. "Trends in Premarital Childbearing: 1930–1994," Census.gov, October 1999, https://www.census.gov/prod/99pubs/p23-197.pdf.

32. Christina Hoff Sommers, "Reconsiderations: Betty Friedan's 'The Feminine Mystique,'" NYSun.com, September 17, 2008, https://www.nysun.com/arts/reconsiderations-betty-friedans-the-feminine/86003/.

33. Betty Friedan, *The Feminist Mystique* (New York: Norton, 2001), 480.

34. Betty Friedan, *It Changed My Life* (Cambridge, Massachusetts: Harvard University Press, 1998), 397

35. Simone de Beauvoir, *The Second Sex* (New York: Vintage Books, 1989), 97, 486.

36. Sue Ellin Browder, "Kinsey's Secret: The Phony Science of the Sexual Revolu-

tion," CrisisMagazine.com, May 28, 2012, https://www.crisismagazine.com/2012
/kinseys-secret-the-phony-science-of-the-sexual-revolution.

37. Caleb Crain, "Alfred Kinsey: Liberator or Pervert?," *New York Times*, October 3, 2004, https://www.nytimes.com/2004/10/03/movies/alfred-kinsey-liberator
-or-pervert.html.

38. James H. Jones, *Alfred C. Kinsey: A Life* (New York: Norton, 1997), 524–25.

39. Herbert Marcuse, *Eros and Civilization* (Boston: Beacon Press, 1974), 5.

40. Ibid., 227–28.

41. Theodor Adorno, *The Authoritarian Personality* (London: Verso, 2019), 727–28.

42. Jeremiah Morelock, "Introduction: The Frankfurt School and Authoritarian
Populism—A Historical Outline," in J. Morelock, ed., *Critical Theory and Authoritarian Populism* (London: University of Westminster Press), xv–xvi.

43. Amity Shlaes, *Great Society: A New History* (New York: Harper, 2019), 98–99.

44. "Cory Booker Claims Catholic Schools Use Religion to 'Justify' Discrimination," *Grabien News*, October 11, 2019, https://news.grabien.com/story-cory-booker
-claims-catholic-schools-use-religion-justify-dis.

45. Michael R. Strain, "Beto O'Rourke's Bad Idea to Punish Conservative Churches,"
Bloomberg.com, October 16, 2019, https://www.bloomberg.com/opinion/articles
/2019-10-16/beto-o-rourke-s-bid-to-end-tax-exemptions-for-churches.

46. Jayme Deerwester, "Chris Pratt Defends Church after Ellen Page Calls It 'Infamously Anti-LGBTQ,'" USAToday.com, February 10, 2019, https://www.usatoday
.com/story/life/people/2019/02/10/ellen-page-calls-out-chris-pratt-over-anti-lgbtq
-church-affiliation/2831293002/.

47. Kory Grow, "Ellie Goulding, Salvation Army Clash Over Gay Rights and Half-
time Performance," RollingStone.com, November 14, 2019, https://www.rolling
stone.com/music/music-news/ellie-goulding-salvation-army-gay-rights-912521/.

48. Timothy Bella, "A Gay Catholic Schoolteacher Was Fired for His Same-Sex
Marriage. Now, He's Suing the Archdiocese," WashingtonPost.com, July 12, 2019,
https://www.washingtonpost.com/nation/2019/07/12/cathedral-high-school-law
suit-archdiocese-indianapolis-joshua-payne-elliott/.

49. "Catholic Charities Pulls Out of Adoptions," WashingtonTimes.com, March 14,
2006, https://www.washingtontimes.com/news/2006/mar/14/20060314-010603
-3657r/.

50. Emma Green, "Even Nuns Aren't Exempt from Obamacare's Birth Control Mandate," TheAtlantic.com, July 14, 2015, https://www.theatlantic.com/politics/archive
/2015/07/obama-beats-the-nuns-on-contraception/398519/.

51. Marc Thiessen, "Hillary Clinton's War on Faith," NYPost.com, October 15, 2016,
https://nypost.com/2016/10/15/hillary-clintons-war-on-faith/.

52. Ed Pilkington, "Obama Angers Midwest Voters with Guns and Religion Remark,"
TheGuardian.com, April 14, 2008, https://www.theguardian.com/world/2008/apr
/14/barackobama.uselections2008.

53. Yuval Levin, "Conservatism in an Age of Alienation," *Modern Age* (Spring 2017),
https://eppc.org/publications/conservatism-in-an-age-of-alienation/.

54. Chris Cillizza, "President Obama's Amazingly Emotional Speech on Gun Control," WashingtonPost.com, January 5, 2016, https://www.washingtonpost.com/news/the-fix/wp/2016/01/05/president-obamas-amazingly-emotional-speech-on-gun-control-annotated/.

55. "'A Human Rights Crisis': US Accused of Failing to Protect Citizens from Gun Violence," TheGuardian.com, September 12, 2018, https://www.theguardian.com/us-news/2018/sep/12/us-gun-control-human-rights-amnesty-international.

56. Kate Sullivan and Eric Bradner, "Beto O'Rourke: 'Hell, Yes, We're Going to Take Your AR-15, Your AK-47,'" CNN.com, September 13, 2019, https://www.cnn.com/2019/09/12/politics/beto-orourke-hell-yes-take-ar-15-ak-47/index.html.

57. "Senator-elect Elizabeth Warren Plans to Advocate for Gun Control, Assault Weapons Ban in US Senate," MassLive.com, December 19, 2012, https://www.masslive.com/politics/2012/12/senator-elect_elizabeth_warren_2.html.

58. Michael Bowman, "Obama Calls for Action on Gun Violence" VOA News, January 14, 2013, https://www.voanews.com/usa/obama-calls-action-gun-violence.

59. Leah Libresco, "I Used to Think Gun Control Was the Answer. My Research Told Me Otherwise," WashingtonPost.com, October 3, 2017, https://www.washingtonpost.com/opinions/i-used-to-think-gun-control-was-the-answer-my-research-told-me-otherwise/2017/10/03/d33edca6-a851-11e7-92d1-58c702d2d975_story.html.

60. John Paul Stevens, "Repeal the Second Amendment," NYTimes.com, March 27, 2018, https://www.nytimes.com/2018/03/27/opinion/john-paul-stevens-repeal-second-amendment.html.

61. Rebecca Savransky, "Connecticut Governor: There's Blood on the NRA's Hands," TheHill.com, March 11, 2018, https://thehill.com/blogs/blog-briefing-room/news/377831-connecticut-governor-theres-blood-on-the-nras-hands.

62. Cillizza, "President Obama's Amazingly Emotional Speech on Gun Control."

63. J. D. Tuccille, "What Will Gun Controllers Do When Americans Ignore an 'Assault Weapons' Ban?," Reason.com, June 21, 2016, https://reason.com/2016/06/21/what-will-gun-controllers-do-when-americ/.

64. David DeGrazia, "Gun Rights Include the Right Not to Be Shot," *Baltimore Sun*, March 15, 2016, https://www.baltimoresun.com/opinion/op-ed/bs-ed-shooting-rights-20160315-story.html.

65. Brad Slager, "CNN Gets Rewarded For, and Brags About, the Disastrous Parkland Town Hall," SunshineStateNews.com, March 21, 2019, http://www.sunshinestatenews.com/story/cnn-gets-rewarded-and-brags-about-disastrous-parkland-town-hall.

66. Associated Press, "Superstore Chain Fred Meyer to Stop Selling Guns and Ammo," LATimes.com, March 17, 2018, https://www.latimes.com/nation/la-na-fred-meyer-guns-20180317-story.html.

67. Kate Gibson, "Kroger to Stop Selling Guns of All Kinds," CBSNews.com, March 19, 2018, https://www.cbsnews.com/news/kroger-to-stop-selling-guns-of-all-kinds/.

68. Alexis de Tocqueville, *Democracy in America*, trans. Harvey C. Mansfield and Delba Winthrop (Chicago: University of Chicago Press, 2000), 290.

69. Ben Shapiro, "The AOC School of Economics," Creators Syndicate, January 24, 2020, https://www.grandforksherald.com/opinion/columns/4878100-Ben-Shapiro-The-AOC-School-of-Economics.

70. Tara Golshan, "Read: Bernie Sanders Defines His Vision for Democratic Socialism in the United States," Vox.com, June 12, 2019, https://www.vox.com/2019/6/12/18663217/bernie-sanders-democratic-socialism-speech-transcript.

71. Edward Conard, *The Upside of Inequality* (New York: Penguin, 2016).

72. Dierdre Nansen McCloskey, "Equality vs. Lifting Up the Poor," *Financial Times*, August 12, 2014, https://www.deirdremccloskey.com/editorials/FTAug2014.php.

73. Doyle McManus, "Most 2020 Democrats Say Capitalism Is a System That Needs Fixing," *Los Angeles Times*, March 20, 2019, https://www.latimes.com/politics/la-na-pol-democrats-socialism-capitalism-20190320-story.html.

74. "Hillary Clinton: Being a Capitalist 'Probably' Hurt Me in Dem Primaries," Real ClearPolitics.com, May 2, 2018, https://www.realclearpolitics.com/video/2018/05/02/hillary_clinton_being_a_capitalist_probably_hurt_me_in_dem_primaries.html.

75. "Noah: For Progressive White People, Being Called Rich Is Equivalent to the 'N-Word,'" Grabien.com, https://grabien.com/story.php?id=265585.

76. William McGurn, "Must Freedom Destroy Itself?," *Wall Street Journal*, June 7, 2019, https://www.wsj.com/articles/must-freedom-destroy-itself-11559944247.

77. Terry Schilling, "How to Regulate Pornography," FirstThings.com, November 2019, https://www.firstthings.com/article/2019/11/how-to-regulate-pornography.

78. Joseph Schumpeter, *Capitalism, Socialism, and Democracy* (New York: Harper Perennial, 2008), 143.

79. Various, "Against the Dead Consensus," FirstThings.com, March 21, 2019, https://www.firstthings.com/web-exclusives/2019/03/against-the-dead-consensus.

80. Senator Marco Rubio, "Catholic Social Doctrine and the Dignity of Work," Catholic University of America, November 5, 2019, https://www.rubio.senate.gov/public/_cache/files/6d09ae19-8df3-4755-b301-795154a68c59/C58480B07D02452574C5DB8D603803EF.final---cua-speech-11.5.19.pdf.

81. Ben Shapiro, "America Needs Virtue Before Prosperity," NationalReview.com, January 8, 2019, https://www.nationalreview.com/2019/01/tucker-carlson-populism-america-needs-virtue-before-prosperity/.

82. Isaac Stanley-Becker, "'She Sounds like Trump at His Best': Tucker Carlson Endorses Elizabeth Warren's Economic Populism," *Washington Post*, June 6, 2019, https://www.washingtonpost.com/nation/2019/06/06/she-sounds-like-trump-his-best-tucker-carlson-endorses-elizabeth-warrens-economic-populism/.

83. Sabrina Tavernise, "Frozen in Place: Americans Are Moving at the Lowest Rate on Record," *New York Times*, November 20, 2019, https://www.nytimes.com/2019/11/20/us/american-workers-moving-states-.html?action=click&module=Latest&pgtype=Homepage.

84. Leigh Buchanan, "American Entrepreneurship Is Actually Vanishing. Here's Why," Inc.com, May 2015, https://www.inc.com/magazine/201505/leigh-buchanan/the-vanishing-startups-in-decline.html.

CHAPTER 5: THE AMERICAN HISTORY

1. Henry Louis Gates, "How Many Slaves Landed in the US?," PBS.org, https://www.pbs.org/wnet/african-americans-many-rivers-to-cross/history/how-many-slaves-landed-in-the-us/.

2. "Emancipation," National Archives (UK), https://www.nationalarchives.gov.uk/pathways/blackhistory/rights/emancipation.htm.

3. Frederick Douglass, "What to the Slave Is the Fourth of July?," TeachingAmerican History.org, July 5, 1852, https://teachingamericanhistory.org/library/document/what-to-the-slave-is-the-fourth-of-july/.

4. Colman Andrews, "These Are the 56 People Who Signed the Declaration of Independence," *USA Today*, July 3, 2019, https://www.usatoday.com/story/money/2019/07/03/july-4th-the-56-people-who-signed-the-declaration-of-independence/39636971/.

5. John Adams, Letter to Robert J. Evans, June 8, 1819, https://founders.archives.gov/documents/Adams/99-02-02-7148.

6. Benjamin Franklin, "Petition from the Pennsylvania Society for the Abolition of Slavery," February 3, 1790, http://www.benjamin-franklin-history.org/petition-from-the-pennsylvania-society-for-the-abolition-of-slavery/.

7. Thomas G. West, *Vindicating the Founders: Race, Sex, Class and Justice in the Origins Of America* (Oxford: Rowman & Littlefield, 1997), 7.

8. James Kirschke, *Gouverneur Morris: Author, Statesman, and Man of the World* (New York: Thomas Dunne Books, 2005), 176.

9. George Washington, Letter to Robert Morris, April 12, 1786, https://founders.archives.gov/documents/Washington/04-04-02-0019.

10. "A Decision to Free His Slaves," MountVernon.org, https://www.mountvernon.org/george-washington/slavery/washingtons-1799-will/.

11. "Jefferson's Attitudes Toward Slavery," Monticello.org, https://www.monticello.org/thomas-jefferson/jefferson-slavery/jefferson-s-attitudes-toward-slavery/.

12. Bill to Prevent the Importation of Slaves, June 16, 1777, https://founders.archives.gov/documents/Jefferson/01-02-02-0019.

13. Vermont Constitution, 1777.

14. An Act for the Gradual Abolition of Slavery, March 1, 1780, http://www.phmc.state.pa.us/portal/communities/documents/1776-1865/abolition-slavery.html.

15. "Massachusetts Constitution and the Abolition of Slavery," Mass.gov, https://www.mass.gov/guides/massachusetts-constitution-and-the-abolition-of-slavery.

16. Abraham Lincoln, Cooper Union Address, February 27, 1860, http://www.abrahamlincolnonline.org/lincoln/speeches/cooper.htm.

17. "Africans in America," Library of Congress, https://www.loc.gov/teachers/classroommaterials/presentationsandactivities/presentations/immigration/african4.html.

18. "Madison Debates," August 8, 1787, https://avalon.law.yale.edu/18th_century/debates_808.asp.

19. Abraham Lincoln, Speech on the Kansas-Nebraska Act at Peoria, Illinois, Octo-

ber 16, 1854, https://teachingamericanhistory.org/library/document/speech-on-the
-kansas-nebraska-act-at-peoria-illinois-abridged/.

20. "Madison Debates," August 25, 1787, https://avalon.law.yale.edu/18th_century
/debates_825.asp.

21. "The Growing New Nation," PBS.org, https://www.pbs.org/wgbh/aia/part3
/map3.html.

22. Thomas Jefferson, "Notes on the State of Virginia, Query XVIII: Manners,"
1781, https://teachingamericanhistory.org/library/document/notes-on-the-state-of
-virginia-query-xviii-manners/.

23. Russell Thornton, "Population History of Native North Americans," in Michael R. Haines and Richard H. Steckel, eds., *A Population History of North America*
(Cambridge: Cambridge University Press, 2000), 24.

24. "Table 1. Population and Area: 1790 to 2000," Census.gov, ftp://ftp.census.gov
/library/publications/2010/compendia/statab/130ed/tables/11s0002.pdf.

25. "Following the Frontier Line, 1790 to 1800," https://www.census.gov/dataviz
/visualizations/001/.

26. Thomas Jefferson, Letter to George Rogers Clark, December 25, 1780, https://
founders.archives.gov/documents/Jefferson/01-04-02-0295.

27. "North America in 1800," NationalGeographic.org, https://www.national
geographic.org/photo/northamerica-colonization-1800/.

28. Thomas Jefferson, Letter to George Rogers Clark, December 25, 1780, https://
founders.archives.gov/documents/Jefferson/01-04-02-0295.

29. "December 2, 1823: Seventh Annual Message (Monroe Doctrine)," University of Virginia Miller Center, https://millercenter.org/the-presidency/presidential
-speeches/december-2-1823-seventh-annual-message-monroe-doctrine.

30. Johnson, *The Birth of the Modern*, 27–28.

31. Ibid., 32–33.

32. Larry Schweikart and Michael Allen, *A Patriot's History of the United States* (New
York: Penguin, 2004), 207–9.

33. Paul Johnson, *The Birth of the Modern* (New York: HarperCollins, 1991), 211.

34. Douglas Bamforth, "Intertribal Warfare," Encyclopedia of the Great Plains,
http://plainshumanities.unl.edu/encyclopedia/doc/egp.war.023.

35. Lawrence H. Keeley, *War Before Civilization* (Oxford: Oxford University Press, 1996).

36. Tocqueville, *Democracy in America*, 391.

37. Ibid., 324.

38. 1860 U.S. Census, https://www2.census.gov/library/publications/decennial/1860
/population/1860a-02.pdf.

39. Alan Greenspan and Adrian Woolridge, *Capitalism in America: A History* (New
York: Penguin, 2018), 77–79.

40. John C. Calhoun, "Speech on the Oregon Bill," June 27, 1848, https://teaching
americanhistory.org/library/document/oregon-bill-speech/.

41. Abraham Lincoln, "House Divided Speech," June 16, 1858, http://www.abraham
lincolnonline.org/lincoln/speeches/house.htm.

42. "The Declaration of the Causes of Seceding States: South Carolina," December 20, 1860, https://www.battlefields.org/learn/primary-sources/declaration-causes-seceding-states#South_Carolina.

43. Constitution of the Confederate States, March 11, 1861, https://avalon.law.yale.edu/19th_century/csa_csa.asp.

44. Henry Louis Gates, "The Truth Behind '40 Acres and a Mule,'" PBS.org, https://www.pbs.org/wnet/african-americans-many-rivers-to-cross/history/the-truth-behind-40-acres-and-a-mule/.

45. Homestead Act (1862), https://www.ourdocuments.gov/doc.php?flashfalse&doc=31.

46. Greenspan and Woolridge, *Capitalism in America*, 77–79.

47. Ibid., 92–96.

48. *Lochner v. New York* (1905), 198 US 45.

49. Daniel Oliver, "James J. Hill: Transforming the American Northwest," FEE.org, July 1, 2001, https://fee.org/articles/james-j-hill-transforming-the-american-northwest/.

50. Schweikart and Allen, *A Patriot's History of the United States*, 459.

51. "From the Archives: President Teddy Roosevelt's New Nationalism Speech," WhiteHouse.gov, August 31, 1910, https://obamawhitehouse.archives.gov/blog/2011/12/06/archives-president-teddy-roosevelts-new-nationalism-speech.

52. Bruce S. Thornton, *Democracy's Dangers and Discontent* (Stanford, CA: Hoover Institution Press, 2014).

53. Lawrence W. Reed, "Of Meat and Myth," Mackinac.org, February 13, 2002, https://www.mackinac.org/4084.

54. Jim Powell, *Bully Boy: The Truth About Theodore Roosevelt's Legacy* (New York: Crown Forum, 2006).

55. Jonah Goldberg, "You Want a More 'Progressive' America? Be Careful What You Wish For," CSMonitor.com, February 5, 2008, https://www.csmonitor.com/Commentary/Opinion/2008/0205/p09s01-coop.html.

56. Woodrow Wilson, *The Political Thought of Woodrow Wilson* (Indianapolis: Bobbs-Merrill, 1965), 192.

57. Judson MacLaury, "A Brief History: The US Department of Labor," DOL.gov, https://www.dol.gov/general/aboutdol/history/dolhistoxford.

58. Geoffrey Wawro, "Everything You Know About How World War I Ended Is Wrong," Time.com, September 26, 2018, https://time.com/5406235/everything-you-know-about-how-world-war-i-ended-is-wrong/.

59. Goldberg, "You Want a More 'Progressive' America?"

60. Woodrow Wilson, "War Message to Congress," April 2, 1917, https://wwi.lib.byu.edu/index.php/Wilson%27s_War_Message_to_Congress.

61. MacLaury, "A Brief History: The US Department of Labor."

62. Calvin Coolidge, "Speech on the 150th Anniversary of the Declaration of Independence," July 5, 1926, https://teachingamericanhistory.org/library/document/speech-on-the-occasion-of-the-one-hundred-and-fiftieth-anniversary-of-the-declaration-of-independence/.

63. Greenspan and Woolridge, *Capitalism in America*, 193–96.

64. Ben Shapiro, "Herbert Hoover's Daughter Hillary," Creators Syndicate, April 30, 2003, https://townhall.com/columnists/benshapiro/2003/04/30/herbert-hoovers-daughter-hillary-n1146318.

65. Steven Horwitz, "Hoover's Economic Policies," EconLib.org, https://www.econlib.org/library/Enc/HooversEconomicPolicies.html.

66. Albin Krebs, "Rexford Tugwell, Roosevelt Aide, Dies," *New York Times*, July 24, 1979, https://www.nytimes.com/1979/07/24/archives/rexford-tugwell-roosevelt-aide-dies-recruited-for-inner-circle-in.html.

67. Steven Horwitz, "Hoover's Economic Policies," EconLib.org, https://www.econlib.org/library/Enc/HooversEconomicPolicies.html.

68. Franklin D. Roosevelt, "Campaign Address," October 14, 1936, https://teachingamericanhistory.org/library/document/campaign-address/.

69. Franklin D. Roosevelt, Re-Nomination Acceptance Speech, July 27, 1936, https://teachingamericanhistory.org/library/document/acceptance-speech-at-the-democratic-national-convention-1936/.

70. Samuel Staley, "FDR Policies Doubled the Length of the Great Depression," Reason.org, November 21, 2008, https://reason.org/commentary/fdr-policies-doubled-the-lengt/.

71. Harold L. Cole and Lee E. Ohanian, "How Government Prolonged the Depression," *Wall Street Journal*, February 2, 2009, https://www.wsj.com/articles/SB123353276749137485.

72. *Biennial Reports of the Chief of Staff of the United States Army to the Secretary of War, 1 July 1939–30 June 1945* (Washington, DC: Center of Military History, 1996), v, https://history.army.mil/html/books/070/70-57/CMH_Pub_70-57.pdf.

73. Winston Churchill, "We Shall Fight on the Beaches," June 4, 1940, https://winstonchurchill.org/resources/speeches/1940-the-finest-hour/we-shall-fight-on-the-beaches/.

74. Franklin D. Roosevelt, Radio Address, December 29, 1940, https://www.mtholyoke.edu/acad/intrel/WorldWar2/arsenal.htm.

75. Carl M. Cannon, "When Churchill, FDR Found Themselves 'in the Same Boat,'" RealClearPolitics.com, December 26, 2018, https://www.realclearpolitics.com/articles/2018/12/26/when_churchill_fdr_found_themselves_in_the_same_boat_139023.html.

76. Greenspan and Woolridge, *Capitalism in America*, 273.

77. "Marshall Plan," History.com, December 16, 2009, https://www.history.com/topics/world-war-ii/marshall-plan-1.

78. "Berlin Airlift," History.com, March 2, 2011, https://www.history.com/topics/cold-war/berlin-airlift.

79. Harry Truman, Statement on the Situation in Korea, June 27, 1950, https://www.docsteach.org/documents/document/truman-statement-korea.

80. Stephen Thernstrom and Abigail Thernstrom, *America in Black and White: One Nation, Indivisible* (New York: Simon & Schuster, 1997), 234.

81. Tobin Grant, "The Great Decline: 60 Years of Religion in One Graph," Religion News.com, January 27, 2014, https://religionnews.com/2014/01/27/great-decline -religion-united-states-one-graph/.

82. Lyndon B. Johnson, "Commencement Address at Howard University: 'To Fulfill These Rights,'" June 4, 1965, https://teachingamericanhistory.org/library/document /commencement-address-at-howard-university-to-fulfill-these-rights/.

83. Amity Shlaes, *Great Society: A New History* (New York: HarperCollins, 2019), 9.

84. Thernstrom and Thernstrom, *America in Black and White*, 234–35.

85. Daniel Patrick Moynihan, "The Negro Family: The Case for National Action," March 1965, https://www.dol.gov/general/aboutdol/history/webid-moynihan.

86. Thernstrom and Thernstrom, *America in Black and White*, 234–35.

87. Greenspan and Woolridge, *Capitalism in America*, 306.

88. Shlaes, *Great Society*, 304.

89. Bryan Burrough, "The Bombings of America That We Forgot," Time.com, September 20, 2016, https://time.com/4501670/bombings-of-america-burrough/.

90. Barry Latzer, *The Rise and Fall of Violent Crime in America* (New York: Encounter Books, 2016).

91. Jimmy Carter, "Crisis of Confidence," July 15, 1979, https://www.pbs.org/wgbh /americanexperience/features/carter-crisis/.

92. Ronald Reagan, First Inaugural Address, January 20, 1981, https://avalon.law .yale.edu/20th_century/reagan1.asp.

93. *Hearings on Military Posture and HR 6495* (Washington, DC: U.S. Government Printing Office, 1980), 94.

94. Gabriel Florit, Kim Soffen, Aaron Steckelberg, and Tim Meko, "40 Years of Budgets Show Shifting National Priorities," WashingtonPost.com, March 17, 2017, https://www.washingtonpost.com/graphics/politics/budget-history/?utm_term =.1336a2b613ba.

95. William J. Clinton, 1996 State of the Union Address, January 23, 1996, https:// clintonwhitehouse4.archives.gov/WH/New/other/sotu.html.

96. Eric Schmitt, "Iraq-Bound Troops Confront Rumsfeld Over Lack of Armor," NYTimes.com, December 8, 2004, https://www.nytimes.com/2004/12/08/inter national/middleeast/iraqbound-troops-confront-rumsfeld-over-lack-of.html.

CHAPTER 6: DISINTEGRATING AMERICAN HISTORY

1. Jefferson Davis, Speech at Boston's Faneuil Hall, October 11, 1858, *The Papers of Jefferson Davis*, https://jeffersondavis.rice.edu/archives/documents/jefferson-davis -speech-boston.

2. Ibid.

3. John C. Calhoun, "Speech on the Oregon Bill," June 27, 1848, https://teaching americanhistory.org/library/document/oregon-bill-speech/.

4. William J. Cooper Jr., *Jefferson Davis, American* (New York: Vintage Books, 2001), 1–8.

5. Jefferson Davis, First Inaugural Address, February 18, 1861, *The Papers of Jefferson*

Davis, https://jeffersondavis.rice.edu/archives/documents/jefferson-davis-first-inau
gural-address.

6. Theodore R. Johnson, "How Conservatives Turned the 'Color-Blind Constitution' Against Racial Progress," TheAtlantic.com, November 19, 2019, https://www
.theatlantic.com/ideas/archive/2019/11/colorblind-constitution/602221/.

7. Eve Fairbanks, "The 'Reasonable' Rebels," WashingtonPost.com, August 29,
2019, https://www.washingtonpost.com/outlook/2019/08/29/conservatives-say-weve
-abandoned-reason-civility-old-south-said-that-too/?arc404=true.

8. David W. Blight, "'A Doubtful Freedom,'" NYBooks.com, January 16, 2020,
https://www.nybooks.com/articles/2020/01/16/fugitive-slaves-doubtful-freedom/.

9. Woodrow Wilson, *Constitutional Government in the United States* (New York: Columbia University Press, 1908), 4–5.

10. Milan Griffes, "The Origin and Development of Carl Becker's Historiography,"
https://www.academia.edu/9101433/The_Origin_and_Development_of_Carl
_Beckers_Historiography.

11. Carl Becker, "Everyman His Own Historian," *American Historical Review* 37, no 2
(1931): 221–36, https://www.historians.org/about-aha-and-membership/aha-history
-and-archives/presidential-addresses/carl-l-becker.

12. James Harvey Robinson, *The Human Comedy as Devised and Directed by Mankind
Itself* (New York: Harper & Brothers, 1937), 21.

13. Burton W. Folsom, "The Founders, the Constitution, and the Historians," FEE
.org, June 11, 2009, https://fee.org/articles/the-founders-the-constitution-and-the
-historians/.

14. Peter Rutkoff and William B. Scott, *New School: A History of the New School for
Social Research* (New York: Free Press, 1986), 6.

15. Folsom, "The Founders, the Constitution, and the Historians."

16. David Waldstreicher, "Foreword" to Staughton Lynd, *Intellectual Origins of American Radicalism* (New York: Cambridge University Press, 1968), xxxi.

17. Eric Foner, *Who Owns History? Rethinking the Past in a Changing World* (New
York: Hill & Wang, 2002), 27–28.

18. George F. Will, "Candidate on a High Horse," *Washington Post*, April 15, 2008,
https://www.washingtonpost.com/wp-dyn/content/article/2008/04/14/AR200
8041402450.html?tid=lk_inline_manual_2.

19. David Plotkinoff, "Zinn's Influential History Textbook Has Problems, Says Stanford Education Expert," Stanford.edu, December 2012, https://news.stanford.edu
/news/2012/december/wineburg-historiography-zinn-122012.html.

20. Mary Grabar, *Debunking Howard Zinn: Exposing the Fake History That Turned a
Generation Against America* (Washington, DC: Regnery History, 2019), 40.

21. Howard Zinn, *A People's History of the United States* (New York: HarperPerennial,
2015), 10–11.

22. Ibid., 59.

23. Ibid., 198.

24. Ibid., 424.

25. Ibid., 430.

26. Sam Wineburg, "Undue Certainty: Where Howard Zinn's *A People's History* Falls Short," *American Educator*, Winter 2012–2013, https://www.aft.org/sites/default /files/periodicals/Wineburg.pdf.

27. "Eric Foner: The Best Antidote to Bad History Is Good History," History News Network, April 11, 2017, https://historynewsnetwork.org/article/165667.

28. Payne Hiraldo, "The Role of Critical Race Theory in Higher Education," *Vermont Connection* 31 (2010): 53–59, https://www.uvm.edu/~vtconn/v31/Hiraldo.pdf.

29. Gloria Ladson-Billings and William F. Tate IV, "Toward a Critical Race Theory of Education," *Teachers College Record* 97, no. 1 (Fall 1995), http://hs.iastate.edu/wp -content/uploads/2011/01/Toward_a_Critical_Race_Theory_of_Education.pdf.

30. "Transcript: Barack Obama's Speech on Race," NPR.org, March 18, 2008, https:// www.npr.org/templates/story/story.php?storyId=88478467.

31. Bill Chappell, "'We Are Not Cured': Obama Discusses Racism in America with Marc Maron," NPR.org, June 22, 2015, https://www.npr.org/sections/thetwo-way /2015/06/22/416476377/we-are-not-cured-obama-discusses-racism-in-america -with-marc-maron.

32. Victor Davis Hanson, "Obama: Transforming America," RealClearPolitics.com, October 1, 2013, https://www.realclearpolitics.com/articles/2013/10/01/obama_trans forming_america_120170.html.

33. Jake Silverstein, "Why We Published the 1619 Project," NYTimes.com, December 20, 2019, https://www.nytimes.com/interactive/2019/12/20/magazine/1619 -intro.html.

34. Adam Serwer, "The Fight Over the 1619 Project Is Not About the Facts," The Atlantic.com, December 23, 2019, theatlantic.com/ideas/archive/2019/12/historians -clash-1619-project/604093/.

35. Elliot Kaufman, "The 1619 Project Gets Schooled," WSJ.com, December 16, 2019, https://www.wsj.com/articles/the-1619-project-gets-schooled-11576540494.

36. Christina Joseph, "'1619 Project' Poised to Reframe Teaching of Slavery. Here's How Educators Are Using the Information, Curriculum," SLJ.com, October 24, 2019, https://www.slj.com/?detailStory=1619-project-poised-to-reframe-teaching-slavery -how-educators-using-information-curriculum.

37. Heather MacDonald, "Ethnic Studies 101: Playing the Victim," City-Journal.org, January 16, 2020, https://www.city-journal.org/lorgia-garcia-pena-harvard-diversity -debate.

38. Ashley Thorne, "The Drive to Put Western Civ Back in the College Curriculum," *New York Post*, March 29, 2016, https://nypost.com/2016/03/29/the-drive-to-put -western-civ-back-in-the-college-curriculum/.

39. Emily Deruy, "The Complicated Process of Adding Diversity to the College Syllabus," TheAtlantic.com, July 29, 2016, https://www.theatlantic.com/education /archive/2016/07/the-complicated-process-of-adding-diversity-to-the-college -syllabus/493643/.

40. Diane Ravitch, "Decline and Fall of Teaching History," *New York Times*, Novem-

ber 17, 1985, https://www.nytimes.com/1985/11/17/magazine/decline-and-fall-of -teaching-history.html.

41. Nikole Hannah-Jones, Twitter, November 21, 2019, https://twitter.com/nhannah jones/status/1197573220037201922.

42. Michael Harriot, "Black History, According to White People," TheRoot.com, January 16, 2020, https://www.theroot.com/black-history-according-to-white-people -1841047480?utm_source=theroot_twitter&utm_medium=socialflow.

43. Ta-Nehisi Coates, "The Case for Reparations," *The Atlantic*, June 2014, https://www .theatlantic.com/magazine/archive/2014/06/the-case-for-reparations/361631/.

44. Zinn, *A People's History of the United States*, 1.

45. Ibid., 21.

46. Grabar, *Debunking Howard Zinn*.

47. Steven Pinker, "A History of Violence," *New Republic*, March 18, 2007, https:// newrepublic.com/article/77728/history-violence.

48. Silverstein, "Why We Published the 1619 Project."

49. Folsom, "The Founders, the Constitution, and the Historians."

50. Tocqueville, *Democracy in America*, 51–52.

51. Ibid., 200.

52. Michelle Alexander, "Injustice on Repeat," NYTimes.com, January 17, 2020, https://www.nytimes.com/2020/01/17/opinion/sunday/michelle-alexander-new -jim-crow.html?action=click&module=Opinion&pgtype=Homepage.

53. Matthew Desmond, "In Order to Understand the Brutality of American Capitalism, You Have to Start on the Plantation," NYTimes.com, August 14, 2019, https:// www.nytimes.com/interactive/2019/08/14/magazine/slavery-capitalism.html.

54. Tocqueville, *Democracy in America*, 331–32.

55. Greg Toppo and Paul Overberg, "After Nearly 100 Years, Great Migration Begins Reversal," USAToday.com, February 2, 2015, https://www.usatoday.com/story /news/nation/2015/02/02/census-great-migration-reversal/21818127/.

56. Scott Sumner, "Ending Slavery Made America Richer," EconLib.org, September 2014, https://www.econlib.org/archives/2014/09/ending_slavery.html.

57. Coleman Hughes, "Black American Culture and the Racial Wealth Gap," Quillette .com, July 19, 2018, https://quillette.com/2018/07/19/black-american-culture-and -the-racial-wealth-gap/.

58. Raj Chetty, Nathaniel Hendren, Maggie Jones, and Sonya R. Porter, "Race and Economic Opportunity in the United States," Equality of Opportunity Project, March 2018, http://www.equality-of-opportunity.org/assets/documents/race_summary.pdf.

59. Thomas Sowell, *Ethnic America: A History* (New York: Basic Books, 1981), 219.

60. "Children in Single-Parent Families by Race in the United States," Annie E. Casey Foundation Kids Count Data Center, https://datacenter.kidscount.org/data /tables/107-children-in-single-parent-families-by#detailed/1/any/false /867,133,38,35,18/10,9,12,1,185,13/432,431.

61. Barack Obama, "Obama's Father's Day Remarks," NYTimes.com, June 15, 2008, https://www.nytimes.com/2008/06/15/us/politics/15text-obama.html.

62. "Marriage and Poverty in the United States: By the Numbers," Heritage Foundation, 2010, https://thf_media.s3.amazonaws.com/2010/pdf/wm2934_bythenumbers.pdf.

63. "Dropout Rates," National Center for Education Statistics, https://nces.ed.gov/fastfacts/display.asp?id=16.

64. Eli Hager, "A Mass Incarceration Mystery," TheMarshallProject.org, December 15, 2017, https://www.themarshallproject.org/2017/12/15/a-mass-incarceration-mystery.

65. Hughes, "Black American Culture and the Racial Wealth Gap."

66. Karma Allen, "Democrats Tackle Racism, Mass Incarceration on Debate Stage," ABCNews.com, September 12, 2019, https://abcnews.go.com/Politics/democrats-tackle-racism-mass-incarceration-debate-stage/story?id=65583080.

67. Ibid.

68. Mary Margaret Ohlson, "FLASHBACK: Buttigieg Said the Founding Fathers 'Did Not Understand That Slavery Was a Bad Thing,'" DailyCaller.com, December 31, 2019, https://dailycaller.com/2019/12/31/pete-buttigieg-founding-fathers-slavery/.

69. Ian Schwartz, "Biden: 'White Man's Culture' Has Got to Change," RealClear Politics.com, March 26, 2019, https://www.realclearpolitics.com/video/2019/03/26/biden_white_mans_culture_english_jurisprudential_culture_has_to_change.html.

CONCLUSION

1. Adam Edelman, "Trump Blasts Cuomo for Saying 'America Was Never That Great,'" NBCNews.com, August 15, 2018, https://www.nbcnews.com/politics/politics-news/n-y-gov-andrew-cuomo-america-was-never-great-n901071.

2. "Holder: America Was Never Great," FreeBeacon.com, March 28, 2019, https://freebeacon.com/politics/holder-america-was-never-great/.

3. Tocqueville, *Democracy in America*, 663.

INDEX

historical revisionism, of
 Disintegrationists, 171–76. *See also*
 American history
Hofstadter, Richard, 174
Holder, Eric, 197
Holmes, Oliver Wendell, 66–67
Homestead Act (1862), 141
Hoover, Herbert, 149
Hugh of Saint-Victor, 5
Hughes, Coleman, 190
Hughes, Donna, 32
human nature
 American philosophy and rights
  inherent in, 7–8, 25–26, 29, 31, 34,
  36, 61, 65
 as both sinful and rational, 31
 Disintegrationists and malleability of,
  xvii, 26–36, 40, 48, 52
Hussein, Saddam, 162

"In Order to Understand the Brutality of
 American Capitalism, You Have to
 Start at the Plantation" (Desmond),
 187–88
income inequality, 110–11, 185
Intellectual Origins of American Radicalism
 (Lynd), 173–74
intersectionality, Disintegrationism and,
 xxi–xxii, 44–45
Iranian revolution, 159
Iraq War, 162–63
Israel, Scott, 105

Jackson, Andrew, 133, 134
Japanese Americans, 190
Jay, John, 127
Jefferson, Thomas, 124
 Declaration of Independence and,
  3, 10
 equality before the law and, 7–8
 reason's value and, 6
 religious freedom and, 69, 71

right to bear arms and, 73
slavery and, 16, 126, 128, 130
westward migration and "Empire of
 Liberty," 131–32
Jewish Americans, 190
Jim Crow laws, 45–46, 140, 155, 187–90
Johnson, Andrew, 140
Johnson, Lyndon Baines, 27, 99, 157, 158,
 160
Johnson, Paul, 133, 134
Johnson, Samuel, 10
Johnson, Theodore, 44
Journal of the Proceedings of the Congress
 (1774), 9
Judeo-Christian morality, natural rights
 and, 4–6, 12
judiciary
 Constitution and, 20–21
 Disintegrationism and, 55–56
 see also Supreme Court

Kansas-Nebraska Act, 137
Kavanaugh, Brett, 45
Keeley, Lawrence, 135
Kennedy, John F., xx–xxi, 119, 154–55
Kimmel, Jimmy, 88
King, Martin Luther, Jr., xxiv, 2, 11, 61,
 187
Kinsey, Alfred, 98
Kirsanow, Peter, 180
Korea, 154
Ku Klux Klan, 68, 80, 84, 140

labor theory of value, Disintegrationism's
 view of economics and, 108–9, 114
Latzer, Barry, 158–59
Lee, Richard Henry, 7
legislatures
 Disintegrationism and, 51, 55–56
 founders and Constitution, 18–20
Leo XIII, Pope, 115
Levin, Yuval, 101

ABOUT THE AUTHOR

Ben Shapiro is editor in chief of the *Daily Wire* and host of *The Ben Shapiro Show*, the top conservative podcast in the nation. His syndicated radio show reaches nearly every top market in the United States. A *New York Times* bestselling author, Shapiro is a graduate of Harvard Law School. His work has been profiled in nearly every major American publication, and he has appeared as the most-requested speaker at conservative events on campuses nationwide.